T0282614

CAMBRIDGE LIBRARY COLLECTION

Books of enduring scholarly value

Earth Sciences

In the nineteenth century, geology emerged as a distinct academic discipline. It pointed the way towards the theory of evolution, as scientists including Gideon Mantell, Adam Sedgwick, Charles Lyell and Roderick Murchison began to use the evidence of minerals, rock formations and fossils to demonstrate that the earth was older by millions of years than the conventional, Bible-based wisdom had supposed. They argued convincingly that the climate, flora and fauna of the distant past could be deduced from geological evidence. Volcanic activity, the formation of mountains, and the action of glaciers and rivers, tides and ocean currents also became better understood. This series includes landmark publications by pioneers of the modern earth sciences, who advanced the scientific understanding of our planet and the processes by which it is constantly re-shaped.

The Wonders of Geology

Gideon Mantell (1790–1852) was an English physician and geologist best known for pioneering the scientific study of dinosaurs. After an apprenticeship to a local surgeon in Sussex, Mantell became a member of the Royal College of Surgeons in 1811. He developed an interest in fossils, and in 1822 his discovery of fossil teeth which he later identified as belonging to an iguana-like creature he named *Iguanadon* spurred research into ancient reptiles. These volumes, first published in 1838, contain a series of eight lectures which describe and explain early principles of geology, stratification and fossil plants and animals to a non-scientific audience. These detailed volumes became Mantell's most popular work, and provide a fascinating view of the study of geology and palaeontology during the early nineteenth century. Volume 2 contains lectures 5–8, discussing marine fossils including animals and corals, fossil plants and volcanic rocks.

Cambridge University Press has long been a pioneer in the reissuing of out-of-print titles from its own backlist, producing digital reprints of books that are still sought after by scholars and students but could not be reprinted economically using traditional technology. The Cambridge Library Collection extends this activity to a wider range of books which are still of importance to researchers and professionals, either for the source material they contain, or as landmarks in the history of their academic discipline.

Drawing from the world-renowned collections in the Cambridge University Library, and guided by the advice of experts in each subject area, Cambridge University Press is using state-of-the-art scanning machines in its own Printing House to capture the content of each book selected for inclusion. The files are processed to give a consistently clear, crisp image, and the books finished to the high quality standard for which the Press is recognised around the world. The latest print-on-demand technology ensures that the books will remain available indefinitely, and that orders for single or multiple copies can quickly be supplied.

The Cambridge Library Collection will bring back to life books of enduring scholarly value (including out-of-copyright works originally issued by other publishers) across a wide range of disciplines in the humanities and social sciences and in science and technology.

The Wonders of Geology

Or, a Familiar Exposition of Geological Phenomena

VOLUME 2

GIDEON ALGERNON MANTELL
EDITED BY G.F. RICHARDSON

CAMBRIDGE
UNIVERSITY PRESS

CAMBRIDGE UNIVERSITY PRESS

Cambridge, New York, Melbourne, Madrid, Cape Town, Singapore,
São Paolo, Delhi, Dubai, Tokyo, Mexico City

Published in the United States of America by Cambridge University Press, New York

www.cambridge.org
Information on this title: www.cambridge.org/9781108021128

This edition first published 1838
This digitally printed version 2010

ISBN 978-1-108-02112-8 Paperback

P¹¹.

ZOOPHYTES.

Eliza. M. Montal, del ᵈ

Rolfe & Fletcher, Litho. 11. Cornhill

THE

WONDERS OF GEOLOGY;

BY

GIDEON MANTELL, LL.D. F.R.S.

AUTHOR OF

THE GEOLOGY OF THE SOUTH EAST OF ENGLAND,
ETC. ETC.

Silver Coins of Edward the First, in ironstone.

" To the natural philosopher there is no natural object unimportant
or trifling : from the least of nature's works he may learn the greatest
lesson."—SIR J. F. W. HERSCHEL.

" We know not a millionth part of the wonders of this beautiful
world."—LEIGH HUNT.

VOL. II.

LONDON:

RELFE AND FLETCHER, CORNHILL.

1838.

THE

WONDERS OF GEOLOGY;

OR,

A FAMILIAR EXPOSITION

OF

GEOLOGICAL PHENOMENA;

BEING THE SUBSTANCE OF

A COURSE OF LECTURES DELIVERED AT BRIGHTON.

BY

GIDEON MANTELL, LL.D. F.R.S.

FELLOW OF THE ROYAL COLLEGE OF SURGEONS ; -
AND OF THE LINNEAN AND GEOLOGICAL SOCIETIES OF LONDON AND CORNWALL
HONORARY MEMBER OF THE PHILOMATHIC SOCIETY OF PARIS;
OF THE ACADEMIES OF NATURAL SCIENCES OF PHILADELPHIA ; AND OF ARTS AND
SCIENCES OF CONNECTICUT ; OF THE GEOLOGICAL SOCIETY OF PENNSYLVANIA ;
OF THE PHILOSOPHICAL INSTITUTION OF BOSTON
OF THE HISTORICAL SOCIETY OF QUEBEC ; AND OF THE PHILOSOPHICAL
SOCIETIES OF YORK, NEWCASTLE, ETC.

———◆———

FROM NOTES TAKEN BY G. F. RICHARDSON,
CURATOR OF THE MANTELLIAN MUSEUM, ETC.

VOL. II.

LONDON:

RELFE AND FLETCHER, CORNHILL.

1838.

TABLE OF CONTENTS.

VOL. II.

___

LECTURE V.

LECTURE VI.

LECTURE VII.

LECTURE VIII.

THE

WONDERS OF GEOLOGY.

—

LECTURE V.

LECTURE V.

1. REMARKS ON THE FAUNA AND FLORA OF THE CHALK.—The examination of the Chalk and Wealden has afforded a striking illustration, not only of the nature of oceanic and river deposits in general, but also of the condition of animated nature at the close of the geological cycle which comprises the secondary formations. It will therefore be interesting in this stage of our inquiry to

note the development of animal and vegetable life
during that epoch. The ocean of the chalk appears
to have possessed the principal existing marine
types of organization; it teemed with dog-fish,
lamna, galeus, and other genera of the universal
family of the shark—with fishes related to the
chimera, salmon, smelt, pike, dory, and ray, together
with many now extinct genera. Nautili and other
cephalopoda abounded, as in our tropical seas; and
the family of echinodermata, or sea-urchins, was
profusely developed : star-fish, encrinites, and other
radiaria ; crustacea allied to the crab, lobster,
shrimp, and prawn; univalve and bivalve mollusca;
all these leading divisions of marine existence inhabited its waters. And although we have proof that
numerous genera now no more, together with
others of excessive rarity in the present seas, then
swarmed in prodigious numbers ; and negative
evidence that the cetacea, as the whale, porpoise,
seal, &c. were not among its inhabitants; yet the
varied forms of animal life whose presence in the
ocean of the Chalk is attested by their fossil remains, unquestionably establish that the sea presented the same general conditions, and bore the
same relation with the atmosphere and light, as at
the present time. The most remarkable peculiarity
in the zoological character of the chalk is presented
in the class of Reptiles. Turtles are the only living
marine species, but the chalk ocean was not only
peopled by *Chelonia*, but also by two or more

enormous saurians, (page 318,) for the mosæsaurus lived in the cretaceous seas; and the ichthyosaurus and plesiosaurus (as will be shown hereafter) were exclusively inhabitants of the deep.

2. ZOOLOGICAL CHARACTER OF THE WEALDEN. —In the Wealden epoch the lakes and rivers were peopled by fishes, mollusca, and crustacea; which, though specifically distinct from the recent, presented the same principal types as now exist in fresh water. The lacustrine and marsh plants, and the palms, tree-ferns, and cycadeæ, constitute a flora offering peculiar generic and specific characters, but according with the usual forms of tropical vegetation. The fresh-water turtles, crocodiles, and wading-birds, are in accordance with the fauna of modern tropical regions; but the colossal reptiles,— the iguanodon, megalosaurus, and hylæosaurus,— whose living analogues must be sought in the pigmy iguanas and monitors, constitute a zoological character altogether at variance with the existing economy of animated nature. Here, then, we have evidence of a country *exclusively inhabited by enormous reptiles;* for although the most delicate plants, leaves, and fruits, the fragile bones of birds, the epidermis, and even ligaments of brittle shells, are found, not a vestige of any mammiferous animal has been discovered. I forbear to comment in this place on this astounding fact, of which the tertiary deposites gave no intimation. We have now approached the *Age of Reptiles*—that geological

epoch in which the earth was peopled with enormous oviparous quadrupeds ; and the sea, and
the bodies of fresh-water also teemed with reptile
forms.

3. Site of the Country of the Iguanodon.—
Before I proceed to the consideration of the secondary formations which are antecedent to the Wealden,
I would offer a few remarks on the question relative
to a difference of climate which our discoveries seem
to imply.   From what has been advanced it is natural to inquire, whether at the period of the Wealden these latitudes enjoyed a tropical temperature ;
whether turtles, crocodiles, and gigantic reptiles,
here flourished amid forests of tree-ferns and palms ;
or if the geographical situation of the country of
the Iguanodon was far distant from the country
now occupied by its spoils?   I shall not attempt
to reply to these interrogations, but content myself
with offering some remarks on the appearance of
transport which the fossils of the Wealden exhibit ;
for I may premise, that the state of the organic
remains does not seem to warrant the assumption
that the reptiles and terrestrial plants, like the
zoophytes, mollusca, and fishes of the chalk, lived
and died on the spots where their remains are
found entombed.   With the exception of the beds
of river shells, cyprides, and the equiseta, (which
naturally affect a marshy soil,) all the remains bear
marks of having been transported by water from a
distance.   But although three-fourths of the bones

are more or less broken and rolled—the teeth detached from the jaws—the vertebræ and bones of the extremities, with but very few exceptions, scattered here and there—the stems of the plants torn to pieces ; and all these relics intermixed with pebbles of quartz, flinty-slate, and jasper, affording evidence that these heterogeneous materials have been subjected to the action of waters; yet it is manifest that the action was fluviatile, not littoral. The pebbles, though smooth, are not rounded into beach or shingle ; they have been worn by the action of streams and torrents, but not by the waves of the ocean.   And when we consider that the ligaments and tendons of the joints, even of the existing lizards, possess such strength and tenacity as to render the separation of their limbs extremely difficult, we cannot conceive that the gigantic limbs of the Iguanodon and Megalosaurus could have been dissevered without great violence, except by the decomposition of their tendons from long maceration ; and if this process were alone the cause, the bones would not be found broken and apart from each other, but in apposition, as in the fishes of the chalk, where even the scales, gills, and fins, preserve their natural position.

The specimen of the hylæosaurus (page 365) throws much light on this subject: many of the vertebræ and ribs are broken and splintered, but the fragments still remain near each other ; the bones are dislocated, yet lie in a situation bearing

some relation to their original state. These circumstances indicate that the carcase of the original
must have suffered injury and mutilation, and that
the dislocated and broken bones were held together
by the muscles and integuments ; in this state the
headless trunk must have floated down the river,
and at length sunk into the mud of the delta, and
formed, as it were, a nucleus around which the
stems and leaves of palms, and tree-ferns accumulated, and river shells became intermingled with the
general mass.

The phenomena here contemplated appear to
admit of but one explanation—that of a considerable period of transport ; the carcases of the large
reptiles must have long been exposed to such an
agency, and the river which flowed through the
country of the Iguanodon, must have had its source
far distant, perhaps thousands of miles, from the
delta which it deposited. The course of that river—
the extent of that delta—the situation of that
country, will probably for ever remain unknown.*
These, as I conceive, are the conclusions which the
facts we have examined substantiate.   I do not,
however, mean to intimate that there are not proofs
of the existence of a higher temperature than at
present, in the northern hemisphere, during the

* The Mississippi flows through twenty degrees of latitude
and seven of longitude, drains a valley 3000 miles long and
nearly 1000 broad, and its delta extends out to sea several hundred miles.

Iguanodon epoch; on the contrary, the trees of Portland appear still to occupy the area in which they grew, and the nautili and polyparia of the chalk are not inhabitants of northern seas; but I will defer the further consideration of this problem, till other phenomena have been submitted to our examination.

4. MEDIAL SECONDARY FORMATIONS.—According to the order of chronological arrangement, (page 178, Pl. III.) I proceed to the description of other groups of the secondary formations, and shall comprise in the present discourse the consideration of, 1st, the *Oolitic*, 2d, the *Liassic*, and 3dly, the *Saliferous Systems*. The entire series consists of thick beds of clay, limestone, sand and sandstone, marls, and conglomerates, alternating with each other, and abounding in marine exuviæ; the whole having manifestly been deposited in the basin of the ocean. The organic remains comprise a prodigious number of zoophytes, crinoidea, shells, cephalopoda, crustacea, fishes, and marine reptiles. Trees, plants, coal, lignite, and other vegetable remains, also occur; and wings of insects, bones of terrestrial reptiles, and of one genus of mammalia. Evidence is thus afforded of the existence of countries clothed with vegetation and tenanted by animals.

5. THE OOLITE, or JURA LIMESTONE.—Limestone composed of an aggregation of small rounded grains or spherules, is called *oolite*, or egg-stone, from its fancied resemblance to clusters of small

eggs, or to the roe of a fish. As this structure,
though not confined to the strata under consideration, very generally prevails in the limestones of
this division of the Secondary, the term *Oolite* is
employed to designate that series of strata, which
on the Continent is called the Jura-limestone, (*Jura
Kalk,*) from the mountain range of which it forms
so essential a character.

As the plan of these Lectures only embraces a
very general and familiar summary of geological
phenomena, it would be irrelevant to enter upon
details, which, however interesting to the geologist, would embarrass the general observer by the
overwhelming mass of facts that would require his
attention. It is, however, necessary to present an
outline of the leading subdivisions of the Oolite, in
which the Lias may also be included; for although
the series of strata designated by these terms are
separated in most artificial arrangements, we shall
find it convenient, for the sake of brevity, to comprise them in one general survey.

6. TABULAR VIEW OF THE OOLITE AND LIAS.
—The following tabular view of the strata, principally derived from the works of Professor Phillips,
will obviate the necessity for detailed descriptions
of the lithological characters of these deposites.

THE OOLITIC SYSTEM.

UPPER OOLITE of Portland, Wilts, Bucks, Berks, &c.

1. *Portland Oolite*—limestone abounding in ammonites, trigoniæ, &c. and other marine exuviæ—green and ferruginous sands—layers of chert.
2. *Kimmeridge clay*—blue clay, with septaria and bands of sandy concretions—marine shells and other organic remains—*ostrea delta.*

MIDDLE OOLITE of Oxford, Bucks, Yorkshire, &c.

1. *Coral-Oolite*, or Coral-rag—limestone composed of corals, with shells and echini.
2. *Oxford Clay;* with septaria and numerous fossils—beds of calcareous grit, called Kelloway-rock, abounding in organic remains.

LOWER OOLITE of Gloucestershire, Oxfordshire, and Northamptonshire.

1. *Cornbrash*—a coarse shelly limestone.
2. *Forest marble*—sand, with concretions of fissile arenaceous limestone — coarse shelly oolite—sand, grit, and blue clay.
3. *Great Oolite*—calcareous oolitic limestone and freestone; reptiles, corals, &c. upper beds shelly. *Stonesfield slate*—land plants, insects, reptiles, mammalia.
4. *Fuller's earth beds*—marls and clays, with fuller's earth — sandy limestones and shells.
5. *Inferior Oolite*—coarse limestone—masses of conglomerated terebratulæ and other shells — ferruginous sand and concretionary blocks of sandy limestone, and shells.

LOWER OOLITE of Brora in Scotland.

1. *Shelly limestones*—alternations of sandstones, shales, and ironstone, with plants.
2. *Ferruginous limestone,* with carbonized wood and shells.
3. *Sandstone and shale*, with *two beds of coal.*

LOWER OOLITE
of the
Yorkshire coast.

1. *Cornbrash.*
2. *Sandstones and clays,* with land plants, *thin coal and shale*—calcareous sandstone and shelly limestone.
3. *Sandstone,* often carbonaceous, with clays, full of plants ; *coal beds* and *ironstone.*
4. *Limestone,* ferruginous and concretionary sand.

THE LIAS.

LIAS
of
Dorsetshire,
Somersetshire,
Northamptonshire,
and
Yorkshire.

1. *Upper Lias shale,* full of saurian remains, belemnites, ammonites, &c. intercalated with the lowermost sand of the Oolite— nodules and beds of limestone.
2. *Lias marlstone* — calcareous, sandy, and ferruginous strata, very rich in terebratulæ and other fossils.
3. *Lower Lias clay and shale*—abounding in shells—*gryphea incurva,* &c.—interlaminations of sands and nodules of limestone.
4. *Lias rock;* a series of laminated limestones, with partings of clay.

This list of the strata, extensive as it appears, notes only the principal distinctions observable in the immense series of deposites, comprising the oolite and lias. The difference observable between the lower beds of the oolites, in the midland counties, and in Yorkshire and Scotland, is a fact of great interest; and the accumulation of vegetable matter in the state of coal, with the remains of terrestrial plants in Yorkshire and Brora, together with the presence of insects, land-plants, and mam-

malia, in the Stonesfield slate, attest the existence
of land, and the action of rivers and currents. The
observations on the nature of oceanic deposites in
the previous lectures, will have prepared you for
the appearance of such anomalies in the beds of
the ancient seas.

7. GEOGRAPHICAL DISTRIBUTION OF THE
OOLITE AND LIAS.—The Oolite (comprehending
in this term the series of strata above enumerated)
forms a striking feature in the physical geography
of the country, from the southern shore near Ex-
mouth to the Yorkshire coast. It constitutes a
table-land of considerable elevation, the highest
points attaining an altitude of 1500 feet, which
extends in a tortuous line through Yorkshire,
Lincolnshire, Northamptonshire, Oxfordshire, Gloucestershire, Somersetshire, to the coast of Dorsetshire. It generally presents a bold escarpment to
the west, and slopes regularly to the east, dividing
the eastern and western drainage of that part of England.* The Lias forms a district that runs parallel
to the escarpment of the Oolite, from beneath which
it emerges, and traverses the country from the Yorkshire coast, near Redcar, to the cliffs at Lyme Regis.
(See *Geological Map of England,* Plate VI.)

On the continent the Oolite appears in Normandy, and its characteristic fossils prevail in the

* Geology of Yorkshire, by John Phillips, Esq. F.R.S., Professor of Geology of King's College.

quarries around Caen ; diverging into several
branches or ranges of hills, it traverses France,
forms the great mass of the Jura mountains, and
constitutes part of the chain of the Alps, where
beds belonging to this group appear, greatly altered
in composition, by causes to which I have already
alluded.

The usual characteristics of the Lias are well
preserved, even where those of the Oolite are so
blended as to render discrimination difficult.   The
Lias of many parts of Germany can scarcely be
distinguished from that of Dorsetshire; and at
Boll, in Wurtemberg, my friend Dr. Jaeger has discovered bones of ichthyosauri, and other peculiar
liassic fossils.* Even in the Himalayas, argillaceous
beds have been found with fossils, which indicate
a close analogy to those of the Lias ; as the specimens before you, collected by Dr. Royle, clearly
manifest.

Certain subdivisions of the Oolite in England predominate in particular localities ; thus, the Oxford
clay prevails in the midland counties, — the grey
rubbly limestone, called cornbrash, at Malmsbury,
Chippenham, &c.,—the forest marble, in Oxfordshire
and Somersetshire,—the great oolite, at Bath,—the
Stonesfield slate, near Woodstock,—and the inferior
oolite, in the Cotteswold hills.

---

* Uber die Fossile Reptilien welche in Wurtemberg aufge-
funden worden sind.   Stutgard, 1828.

The upper beds of the Oolite on the Continent, are the lithographic slates of Pappenheim, Solenhofen, Monheim, &c., which abound in flying reptiles, insects, crustacea, and marine shells. The Portland rock, which in England is the uppermost division of the series, lies beneath the beds of fresh-water limestone, containing the petrified forest, and teems with ammonites and shells, particularly *trigoniæ*, (Tab. 50, fig. 1,) of which there are several species in this formation. The Kimmeridge clay forms the base of the Isle of Portland, and contains, with numerous other marine shells, a flat species of oyster, *ostrea deltoidea*, so named from its peculiar shape; this oyster is characteristic of the bed. The Coral-rag, or coralline oolite, is literally a petrified coral reef; it is a coarse limestone, the nucleus of which is formed of madrepores, astreæ, and other stony corals, shells, and echini; sand, and pebbles fill up the interstices; the whole being consolidated by calcareous and silicious infiltrations. So obvious is the coralline structure, that the most incurious observer, travelling through the districts where the coral rag abounds, can scarcely fail to remark the blocks of corals which every where meet the eye. From the quarries near Faringdon, in Berks, I have collected, in the course of a few hours, hundreds of specimens of corals, shells, and echini; of the latter, the beautiful, tuberculated cidaris, popularly called *fairy's nightcap*, and its spines, occur in great perfection.

(Tab. 50, fig. 2.)*   The quarries of Calne, in Wiltshire, are particularly rich in these fossils.

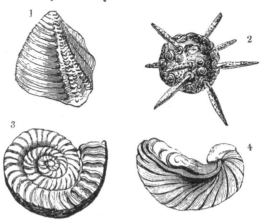

TAB. 50.—FOSSILS OF THE OOLITE AND LIAS.

Fig. 1. *Trigonia costata.*   2. *Cidaris with spines.*   3. *Ammonites*
Walcotii.   4. *Gryphea incurva.*

The relative position of these divisions of the
Oolite with the cretaceous group, is seen in a section

* The heights around Faringdon are generally capped with
green sand, reposing on coral rag. Stanford pit, three miles
south-east of Faringdon, contains:—1. Uppermost, Coral rag, 3½
feet; 2. Limestone, with immense numbers of shells, 4½ feet;
3. Sand, 3 feet; 4. Clay. These beds contain trigoniæ, gervilliæ,
terebratulæ, ostreæ, belemnites, and ammonites: in a slab of
coarse sandy limestone, four feet square, I counted above fifty
gervilliæ, and many trigoniæ. Between Watchfield and Shriven-
ham the coral-rag is seen in openings on the road-side.

near Devizes, in Wiltshire, where the strata appear
in the following order, (see Plate V. fig. II.):—
1. Chalk; 2. Glauconite; 3. Galt; 4. Shanklin
sand; 5. Kimmeridge clay ; 6. Coral-rag ; 7. Oxford
clay. South of Malmsbury, a continuation of the
series to the Lias may be observed, (Plate V.
fig. IV.): — 1. Oxford clay; 2. Great Oolite; 3.
Inferior Oolite ; 4. Lias. On the continent the
coral-rag is largely developed; and near Schaffhausen, and at Muggendorf, the Oolite abounds
in stony polyparia.

8. STONESFIELD SLATE.—I have already stated
that the zoological character of the Oolite and Lias
is decidedly marine; the interspersions of fresh-
water and terrestrial sedimentary deposites, having
been produced by the accumulation of materials
brought down by streams and rivers into the sea,
and transported by currents to a distant part of the
oceanic basin. Unlike the organic remains of the
Wealden, the terrestrial and fresh water productions
are mingled with marine shells, plants, and fishes;
thus, while the Chalk exhibits the bed of a deep
sea without any intermixture of the land or fresh-
water ; and the Wealden a delta in which no marine
exuviæ are imbedded ; the intercalations in the
Oolitic series present a combination of these phenomena, of which the Stonesfield strata afford an
interesting illustration.

Stonesfield, a small village near Woodstock, about
twelve miles north-west of Oxford, has long been

celebrated for the fossil productions of its slaty limestone, the teeth of a large reptile (*Megalosaurus*),
fishes, and other remains, having been described
and figured by Lhwyd, a century ago.   In the
excellent work of Messrs. Conybeare and Phillips,
which every one must regret is not continued and
completed by the highly-gifted author who survives,
there is a description of the Stonesfield strata, and
a brief enumeration of the fossils which they contain.

On my discovery of the fresh-water character of
the strata of Tilgate Forest, I was led to institute a
comparison between the fossils of the Wealden and
those of Stonesfield, and the result appeared in my
first work on the Geology of Sussex.*   A valuable
memoir by Dr. Buckland, on the Megalosaurus,
again drew attention to these interesting deposites,
and hopes were entertained, that this distinguished
philosopher would follow up his investigations, and
give a full description of the organic remains.

The Stonesfield strata have been ascertained, by
Mr. Lonsdale, to belong to the lower division of the
great Oolite; the following account, by Dr. Fitton,†
explains the circumstances under which they occur.
" In crossing the country from Oxford to Stonesfield, the Oxford clay, with its characteristic fossils,
is first observed ; this is succeeded by the Cornbrash, the uppermost stratum of the great Oolite

* Illustrations of the Geology of Sussex, page 37.
† Zoological Journal, Vol. III. page 416.

group, which is seen beneath the clay in several quarries on the road-side between Woodstock and Blenheim. The village of Stonesfield is situated on the brow of a valley, both sides of which are deeply excavated by the shafts and galleries that have been constructed for the extraction of the slate. The beds that supply the stone are at a depth of about fifty feet below the summit, and are worked by shafts. The upper twenty-five feet of strata are of clays alternating with calcareous stone; the lower, of fine-grained oolitic limestone, with numerous casts of shells." From the bottom of the shaft, a drift or horizontal excavation is made around, extending as far as safety will permit; the beds above being supported by piles of the less valuable materials. The strata thus worked do not exceed six feet in thickness; they consist of rubbly stone, with sand imbedding large concretional masses of fine sandy grit, which, by exposure to the frost, admits of separation into thin slates. The resemblance of this calciferous grit to that of Tilgate Forest is most striking; and when breaking it, and perceiving here and there teeth of crocodiles and other reptiles like those of the Wealden, I could have fancied myself sporting in my own geological manor of Tilgate Forest, but for the trigoniæ and other marine shells, and the oolitic structure which every where prevailed. The grit, like that of Sussex, passes into a conglomerate, formed of smooth rounded pebbles, cemented together by oolite; beds of sands, clay, and friable, slaty sandstone, intervene between the layers of the oolitic, calciferous rock. Grits, similar to those of Stonesfield, occur at Wittering and Collyweston, associated with compact limestone and beds of oolite; they contain ferns and other terrestrial plants, and marine shells.

9. ORGANIC REMAINS OF THE STONESFIELD SLATE.—The fossils of Stonesfield, although of so highly interesting a character, have hitherto been very imperfectly investigated. The vegetable re-mains consist of marine plants referrible to fuci; of palms, tree-ferns, and many species of sphenop-teris; plants allied to the zamia and cycas; and a genus of liliaceæ, named *Bucklandia;* seed-vessels, leaves, and stems, of several genera of coniferæ; and of gigantic seeds and grasses. I am not aware that the shells differ from those of the other oolitic strata; a small trigonia (*Trigonia impressa*) is abundant. The bones and teeth of the gigantic reptile related to the monitor,* which I have men-tioned as occurring in the Tilgate grit; teeth and bones of crocodiles, bones and plates of turtles, bones of pterodactyles, or flying-lizards, and other osseous remains, apparently of saurians, present a remarkable correspondence with the fossils of the Wealden. The teeth, scales, fin-bones and rays of fishes, are similar to those contained in

* On the Megalosaurus: Transactions of the Geological Society. 1824,

other beds of the oolite; the round hemispherical teeth of fish allied to the *Lepidotus* of Tilgate Forest are every where in profusion.

10. DIDELPHIS, or OPOSSUM OF STONESFIELD.—But in addition to the extraordinary remains I have enumerated, the Stonesfield slate has yielded to the geologist one of the most precious relics of the past

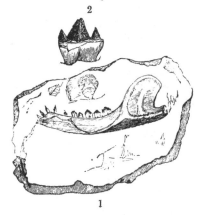

TAB. 51.—LOWER JAW OF AN OPOSSUM, FROM STONESFIELD.
Didelphis Bucklandii.
(*From W. J. Broderip, Esq. F.R.S.*)*
Fig. 1. *The Jaw of its natural size.* 2. *Second molar tooth, magnified six times.*

ages of the globe — the only known example of mammalian remains in the secondary formations;

* Zoological Journal : " Observations on the Jaw of a fossil mammiferous animal in the Stonesfield slate," by W. J. Broderip, Esq.

a fact, standing in this respect in the same rank with the discovery of birds in the Wealden, but of still greater interest, since it carries back the existence of the higher vertebrated animals to a period far more remote. Four specimens of jaws have been discovered, which Mr. Broderip refers to three different species, if not genera. The true affinities of the fossils were determined by Baron Cuvier.* Mr. Broderip's specimen (Tab. 51,) consists of the right half of a lower jaw, the inside of which is exposed; it has seven grinders; one canine, and three incisor teeth; there is a vacancy for a fourth incisor, and thus the number would correspond with the dentition of the recent Didelphis. This fossil is in an admirable state of preservation, and the piece of slate in which it is imbedded has numerous casts of the *Trigonia impressa*, so abundant in the Stonesfield slate.

11. WEALDEN AND STONESFIELD FOSSILS.—A comparative view of the organic remains of the Wealden and the Stonesfield slate, exhibits a striking analogy, in the zoological characters of these deposites.

Grit of Tilgate Forest.	*Stonesfield Slate.*
Wood, in the state of a reddish brown friable mass.	Wood.
Equiseta.	Fuci.
Sphenopteris, Lonchopteris, and other ferns.	Sphenopteris, Tæniopteris, and other ferns.

---

* On the Megalosaurus of Stonesfield: Geological Transactions, Vol. I. page 393. Second Series.

Grit of Tilgate Forest.	*Stonesfield Slate.*
Cycas, or Zamia.	Cycas, or Zamia.
Liliacea.	Liliacea.
Clathraria Lyellii.	Arborescent ferns.
Coniferæ.	Coniferæ.
Seed-vessels—undetermined.	Seed-vessels—undetermined.
Fresh-water Shells.	*Marine Shells.*
Cypris—*Fresh-water* crustacea.	Astacus—*Marine crustacea.*
No insects have been discovered.	*Insects—Coleoptera.*
Fishes of the genera Hybodus, Ptychodus, &c.	Ptychodus, Hybodus, and other fishes.
Lepidotus.	Lepidotus ?
Marine and fresh-water turtles.	Turtles.
Plesiosaurus.	Plesiosaurus.
Pterodactyles.	Pterodactyles.
Crocodiles.	Crocodiles.
Megalosaurus.	Megalosaurus.
Iguanodon, Hylæosaurus, and other reptiles.	Other reptiles.
Bird—Ardea.	*Mammalia—Didelphis.*

The remains of *Cetacea* do not occur in either of these deposites.

From this tabular view, we perceive that the flora and fauna of the Wealden agree in every essential character with those of Stonesfield; the mollusca denote the respective conditions in which the accumulation of the strata took place—the Wealden in the embouchure of a river—the Stonesfield beds in the basin of a deep sea.

12. LITHOGRAPHIC SLATES OF PAPPENHEIM, SOLENHOFEN, AND MONHEIM.—In the Stonesfield and Collyweston grit and shale, we have examples of the occurrence of land animals and plants in the

lowermost division of the oolitic system.   In Germany, the uppermost part of the group contains shales,
and layers of fine-grained, fissile limestone, employed
in lithography, and which afford an assemblage of
organic remains of surpassing interest.   These
deposites are found in that prolongation of the chain
of the Jura which, after the fall of the Rhine at
Schaffhausen, passes into Germany along the borders of the Maine, and near to Cobourg.   The
quarries are situated on the sides of the valley of
the Altmuhl, a tributary of the Danube, which
extends by Pappenheim and Aichsted.   This valley
presents a precipitous escarpment, which is composed of, 1. The uppermost part; calcareous schist,
containing in abundance, fishes, crustacea, asteriæ
and reptiles, with a few small ammonites and bivalve
shells. 2. A magnesian limestone. 3. Limestone
of a greyish white, abounding in ammonites ;
and 4. A brown, or grey sandstone, of a fine
grain, which constitutes the base of the hills of
that district.   The most celebrated quarry of the
calcareous schists, is that of Solenhofen, in the
valley of the Altmuhl, near Pappenheim.*   These
quarries have long been known to contain organic
remains of great beauty and interest.   Some of the
crustacea are so perfect, that the most delicate filaments remain ; as may be seen in this fossil prawn
(*palæmon spinipes*) from Pappenheim. (Tab. 26,

* Oss. Foss. Tom. V.

page 219.*) A saurian, about three feet in length, allied to the crocodile, and numerous flying reptiles, have been found at Solenhofen. Mr. Lyell states† that Count Munster has collected, "seven species of pterodactyles, six saurians, three tortoises, sixty species of fish, forty-six of crustacea, and twenty-six of insects. The number of testacea is comparatively small, as also of plants, which are all marine." The cabinet of Mrs. Murchison contains a beautiful libellula, or dragon-fly, from Solenhofen.‡ M. De la Beche remarks, that the fact of the greatest number of fossil insects yet noticed in the oolite, having been found where the remains of the pterodactyles also occur, seems to establish a connexion between these creatures, not merely accidental; and that it is probable the whole of the deposites of this local group of the Jura limestones, may have been effected on a coast where the water was not deep, and on the shores of which the flying reptiles chased their insect prey. The same geologist considers it probable that these lithographic limestones may have been deposited contemporaneously with the Wealden.

13. COAL OF THE OOLITE.—In the tertiary system of Provence, we noticed the occurrence of beds of coal and carboniferous strata, with limestone containing fresh-water shells and crustacea (page

---

* This tablet and its description have been inserted by mistake in the third Lecture, p. 219.

† Principles of Geology, Vol. IV. p. 289.

‡ Geological Manual, p. 372.

228) ; and in the lacustrine deposites of the Rhine, accumulations of brown coal, or lignite (page 251). In the Wealden, lignite was also observed ; and in some localities (as Bexhill) in such abundance, and associated with shales and laminated sandstones, so much resembling the ancient carboniferous beds, as to have led to an expensive and abortive search for coal. The lower division of the oolite in York-shire, and in Scotland, contains coal formations ; and as these exhibit the transmutation of vegetable sub-stances into a carbonaceous mass, under circum-stances widely different from the examples we have hitherto noticed, I will offer a few remarks on these deposites. Professor Phillips, who has so ably in-vestigated the geology of Yorkshire,* has given a lucid description of the carboniferous strata of the oolite,† and Mr. Murchison, of those of Brora, in Sutherland. The tabular arrangement of the oolitic system (page 385) shows the succession of the deposites in Yorkshire, and Brora.

In the district north of the Humber, the lower oolite assumes a new character : instead of finding beneath the cornbrash, the forest marble and great oolite-beds of sandstone, shale and carbonaceous matter are interpolated above the sand which covers the lias. Proceeding northwards, these strata rapidly increase in thickness, and the carbonaceous layers gradually become concentrated into a stratum

---

* The Geology of Yorkshire.
† Ency. Metrop. art. Geology.

of coal, which, though never exceeding sixteen inches in thickness, is, from local circumstances, of considerable value. These strata assume the appearance of a true coal-field, with subordinate beds of coarse, shelly limestone. The fossil plants which accompany the coal-seams and sandstones, occur also in the calcareous slates and limestones, both on the Yorkshire coast, and at Brandsby. No marine exuviæ have yet been found in the coal grits or shales, with the exception of some bivalves. Along the coast under Gristhorp cliffs, a seam of shale, but a few inches in thickness, may be traced for miles; and, from its abounding in leaves of ferns, equiseta, cycadeæ, and of a great many other plants, it is chiselled out by collectors, to obtain specimens. The beauty and variety of these fossil plants are shown in this interesting and extensive series, presented to me by Dr. Peter Murray, and Mr. Williamson, of Scarborough. "Here," observes Professor Phillips, "we have truly a coal field of the oolitic era, produced by the interposition of vast quantities of sedimentary deposites, brought down by floods from the land, between the more exclusively marine strata of the ordinary oolitic type."

Mr. Murchison, one of our most distinguished and indefatigable geologists, has ascertained that this oolitic carboniferous system extends yet farther northward; and at Brora, and other parts of Sutherland, and on the western coast of Scotland, contains beds of coal of considerable extent. At Brora the sandstones and shales acquire a great thickness, and frequently alternate with layers of plants and beds of coal, from a few inches to nearly four feet in thickness. On the north-east coast of the Isle of Skye, shales and sandstones, with impressions and remains of plants, form an extensive series of deposites above the lias shale.

14. GEOGRAPHICAL DISTRIBUTION OF THE LIAS.—I have stated, in general terms, that the lias in England extends along the western escarpment of the oolite, forming a district which presents an exceedingly variable surface, occasioned by the disruptions of the strata, and subsequent denudations. Its course and extent, from Yorkshire to the Dorsetshire coast, are admirably described by Mr. Conybeare,* from whose work the following abstract is derived.

The Lias, from its northernmost limits on the Yorkshire coast, where it underlies the strata of the eastern moor lands, passes to the south of Whitby and to the east of York, and crosses the Humber, near the junction of the Trent and Ouse; stretching onward beneath the low oolitic range of Lincolnshire, it extends to the Wold hills, on the borders of Nottingham and Lincoln, and the celebrated quarries of Barrow-upon-Soar; whence it continues, accompanying the escarpment of the inferior and great oolite, through Nottingham, Warwick, and Gloucester. Its whole course, to within a few miles south of Gloucester,

* Outlines of the Geology of England and Wales, p. 261.

farther information on this subject, and hasten to notice some of the most interesting local deposites and organic remains of this formation.

20. THE CHELTENHAM WATERS.— The town of Cheltenham is built on the lias, beneath which, but at a very great depth, lies the red marl, the grand depository of the rock-salt and brine-springs of England. The celebrated mineral waters of Cheltenham flow up through the lias, but have their origin in the red marl beds upon which it rests; hence they derive their saliferous ingredients, and undergo various modifications in their passage through the lias. From the analyses of these waters, it appears " that their principal constituents are the chloride of sodium (muriate of soda) or sea-salt, and the sulphates of soda and magnesia. Sulphate of lime, oxide of iron, and chloride of magnesium are present in some wells only, and in much smaller quantities. Besides these ingredients, *iodine* and *bromine* have been detected by Dr. Daubeny, who was desirous of ascertaining whether these two active principles, which the French chemists had recently discovered in modern marine productions, did not also exist in mineral salt-waters, issuing from strata that were formerly beneath the sea. The red marl is the source whence the waters derive their saline properties, but, as the springs pass through the lias marls, which are full of iron pyrites or sulphate of iron, certain chemical changes take place, and hence the

waters derive their celebrated medicinal qualities.
From the decomposition of the sulphate of iron
which take places, a vast quantity of sulphuric acid
must be generated, which, reacting on the different
bases of magnesia, lime, &c. forms those sulphates
so prevalent in the higher or pyritous beds of the
lias, the oxide of iron being at the same time more
or less completely separated. By this means the
mineral waters, which are probably mere brine-
springs at the greatest depths, acquire additional
and more valuable properties as they ascend to the
places from whence they flow. At the same time
it must be borne in mind that fresh water is per-
petually falling from the atmosphere upon the
surface of the lias clay, and more or less perco-
lating through its uppermost strata."*

21. ROCK-SALT AND BRINE-SPRINGS.—Brine-
springs, emanating from water flowing through
subterranean deposites of salt, occur in the great
plains of the red sandstone of Cheshire. Droitwich,
which is situated nearly in the centre of the county,
has long been celebrated for the manufacture of
salt from its brine-springs, which appear to be inexhaustible. It is probable that the manufacture
of salt is coeval with the town itself; but it was not
till the year 1725, that the strong brine for which
it is now famous was discovered; the purity of

* Outline of the Geology of the Neighbourhood of Cheltenham, by R. I. Murchison, Esq. F.R.S. Cheltenham, 1834.

this brine is considered superior to that of any other, and the quantity of salt produced amounts to about 700,000 bushels yearly. At a distance of from thirty to forty feet below the surface is a hard bed of gypsum, which is generally about 150 feet thick : through this a small hole is bored to the river of brine, which is in depth about twenty-two inches, and beneath which is a hard rock of salt. The brine rises rapidly through the aperture, and is pumped into a capacious reservoir, whence it is conveyed into iron boilers for evaporation ; it is supposed to be stronger than any other in the kingdom, and contains above one-fourth part its weight of salt.

The depositories of salt do not however extend over the strata in a connected bed, but occupy limited areas. The saliferous strata of Northwich present the following series :—

		Feet.
1.	Uppermost calcareous marl	15
2.	Red and blue clays	120
3.	Bed of rock salt	75
4.	Clay, with veins of rock salt	31
5.	Second bed of rock salt	110

The origin of these extensive beds of pure salt has not yet been satisfactorily explained, for if we suppose them to have arisen from a mere evaporation of sea-water, it is difficult to account for the absence of all extraneous matter ; it is more probable that their origin may in some measure be due to

igneous action, as chloride of sodium is one of the products of volcanic emanations. The occurrence of the
two most powerful acids, sulphuric (in the gypsum,
or sulphate of lime,) and the muriatic (in the salt),
so largely associated together, is a fact which, in a
more advanced state of chemical knowledge, may
probably throw light on this question.*

22. MAGNESIAN LIMESTONE, OR ZECHSTEIN.—
The magnesian limestone is generally of a light
fawn or yellow colour, and in some parts of a
crystalline, in others of a concretionary character.
In many places, and particularly in the quarries
around Sunderland, it presents a beautiful example
of spheroidal structure, evidently superinduced on
stratified detritus *after its deposition ;* for the laminæ
traverse the globular masses uninterruptedly, as in
the grit of Tilgate Forest, and appear to have been
occasioned by a slow chemical segregation of the
materials. In chalk the same structure sometimes
prevails, as in these examples from Preston chalk-
pit, near Brighton, discovered by Mr. Walter
Mantell. These clusters of spheroids, from Sunderland, exhibit the principal varieties ; some of
them partake so much of the appearance of organic
remains as to have been mistaken for fossils. The
limestone is commonly traversed by veins or strings
of carbonate of lime, and occasionally incloses hollow
geodes of calcareous spar, with sulphate of strontian

* Bakewell's Geology, p. 251.   Lyell's Principles of Geology, vol. ii. p. 17.

or barytes. Galena, sulphuret of zinc, and carbonate of copper, occasionally occur. At Mansfeld,
in Germany, beds of slate, abounding in copper
(Keuper schist), and containing fossil fishes of a
peculiar character, are intercalated in this rock.

23. CONGLOMERATES OF THE UPPER RED
SANDSTONE.—The conglomerates of this formation
are chiefly composed of materials derived from the
disintegration of the more ancient rocks; fragments
and pebbles of slate, quartz rock, granite, porphyry,
&c.; even the fine silicious sandstones have a large
proportion of the detritus of other beds. It would
therefore appear that the sea which deposited the
saliferous group, was bounded by the rocks of whose
ruins it is composed; in like manner as the existence
of beaches of flint-pebbles evinces the destruction of
former chalk-cliffs. The rock on which Nottingham
Castle is built, is a conglomerate containing pebbles
of quartz and primary rocks.*

But the most interesting beds of these conglomerates, or breccias, in this country, are those
in which eruptions of lava appear to have been
thrown into the ocean of the New red sand,
and to have cemented together the water-worn
materials, so as to form a *trap conglomerate;* such
at least seems the origin of the amygdaloidal trap,†
as it is termed, in the vicinity of Exeter. A few
miles to the south of that city, masses of a rock of

* Bakewell's Geol. p. 237.   † Geol. of Engl. & Wales, p. 294.

E E

this kind are interposed between beds of sandstone; the general appearance of the rock is that of a granular mass, somewhat loosely compacted, of a purplish-brown colour, interspersed with minute portions of calcareous spar, mica, and indurated clay tinged by copper or manganese. It is full of small cells, which are filled or lined with manganese, calc-spar, or jasper; a structure termed amygdaloidal (almond-like) in geology; the substance of the rock is an earthy felspar.

24. ORGANIC REMAINS OF THE SALIFEROUS GROUP.—The New red sandstone formation presents a remarkable contrast, in the paucity of organic remains, with the oolite and lias; for while the latter teem with marine exuviæ, and the bones of reptiles, the former, except in a few localities, is destitute of fossils; a proof that the strata were accumulated under circumstances unfavourable to the preservation of animals and vegetables.

Six or more species of fuci have been collected at Mansfeld, and are figured by M. Adolphe Brongniart, in his " Fossil Vegetables ;" in the entire series, twenty-three species of ferns or other cryptogamia, and seventeen of coniferæ and of other families, have been identified.

The polyparia, or corals, which are in such profusion in the oolite, yield but six or seven species; and the radiaria only the same number. A remarkably beautiful species of the crinoidea, or lily-shaped animals, occurs, however, in the Muschel-kalk exclusively; it has not been discovered in England. The specimen in my museum, which belonged to

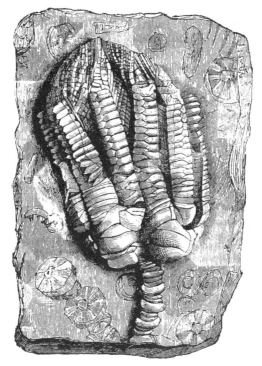

TAB. 52.—THE LILY ENCRINITE, FROM BRUNSWICK.
Encrinus monileformis.

the late Mr. Parkinson, is in great perfection, and admirably displays the structure of the original; in

E E 2

the lecture on zoophytes, (p. 508,) the nature of the singular family to which it belongs will be explained.

Ammonites, nautili, belemnites, and about one hundred species of other mollusca, are specified as having been collected in the various strata of this formation. Among these, two genera, the species of which were formerly referred to anomia and terebratula, first appear, namely, *Producta* and *Spirifera;* in the older strata, which we shall here-after examine, various species of these shells will be found to swarm in the rocks.

25. THE SPIRIFERÆ.—I will offer a few re-marks on the Spiriferæ in this place, that I may introduce the interesting account of the structure of the recent analogues, by my friend Professor Owen, of the Royal College of Surgeons. The small subglobular bivalves, so abundant in the chalk, (*terebratulæ*) are sometimes found empty, and if the valves be carefully separated, two curious ap-pendages are seen projecting from the hinge into the interior of the shell; these processes are the internal skeleton for the support of the organs of respiration. In the spiriferæ (Tab. 53, figs. 1, 6, 9, 10, 11,) there are two spiral appendages (hence the name) which are closely coiled, and are often, like the substance of the shell itself, changed into calcareous spar (see figs. 1, 9); in specimens where the shell is removed, these organs may be seen in their original situation. The following descrip-tion of a recent animal of the same family, a native

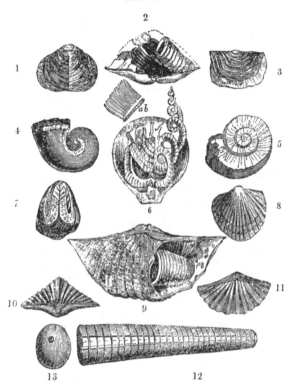

TAB. 53.—FOSSILS OF THE NEW RED SANDSTONE, TRANSITION
SERIES, &c.

Fig. 1. *Producta punctata.* 2. *Spirifera trigonalis, with spiral processes,
from the mountain limestone.* 3. *Producta depressa.* 4. *Bellerophon
cornu-arietis.* 5. *Euomphalus pentangulatus. Carb. limestone.* 6.
Terebratula psittacea, showing the spiral arms. ab. *Cilia of the arms
magnified.* 7. *Pentremites ellipticus. Carbon. formation.* 8. *Terebratula affinis.* 9. *Spirifera trigonalis.* 10. *Spirifera triangularis.* 11. *Spirifera octoplicata.* 12. *Orthoceratite.* 13. *Septum of orthoceratite.*

of the South Seas, will explain the nature of this structure :—

" The loop-like processes observable in the interior of the shells of many of the fossil terebratulæ, are the internal skeleton, and are for the attachment of the muscular stems of the arms. In *Terebratula psittacea*, a recent species, (Tab. 53, fig. 6,) two spiral arms, fringed at their outer margins, are seen to arise from these processes ; these arms are quite free, except at their origins ; when unfolded they are twice as long as the shell ; and in a state of contraction are disposed in six or seven spiral gyrations, which decrease toward their extremities. The mechanism by which the arms is extended is most beautiful and simple ; the stems are hollow from one end to the other, and are filled with fluid, which being acted upon by the spirally disposed muscles composing the walls of the canal, is forcibly injected towards the extremity of the arms, which is thus unfolded and protruded. The respiration, as well as the nutrition, of animals living beneath a pressure of from sixty to ninety fathoms of sea-water, are subjects of peculiar interest, and prepare the mind to contemplate, with less surprise, the wonderful complexity exhibited in the minutest parts of the frame of these diminutive creatures. In the stillness pervading these abysses, they can only maintain existence by exciting a perpetual current around them, in order to dissipate the water already loaded with their effete particles, and bring within the reach of their prehensile organs the animalculæ adapted for their support.

" The spiral disposition of the arms is common to the whole of the brachiopodous genera, whose organization has hitherto been examined ; and it is therefore probable, that in that remarkable genus *Spirifer*, the entire *brachia* were similarly disposed, and that the internal, calcareous, spiral appendages were their supports. If indeed the *brachia* of *Ter. psittacea* had been so sustained, this species would have presented in a fossil state, an internal structure very similar to that of *Spirifer.*"

Tab. 53. fig. 6, page 421. *Terebratula psittacea*, with the perforated valve and lobe of the mantle removed to show the fringed brachia or labial arms, one of which has been artificially unfolded. (Zool. Trans. Pl. XXII. fig. 14.)

Fig. *a b*. A portion of the ciliary fringe of one of the spiral arms magnified.*

Of fishes, fifteen species, of a genus (*Palæothrissum*) supposed to be peculiar to this formation, have been discovered, and are of frequent occurrence at Mansfeld; the same species of ichthyolites have been found in England.

Nine species of reptiles are enumerated by authors; of these some are referrible to the crocodile, others to the ichthyosaurus and plesiosaurus, and a few belong to new or unknown genera; one of these is described by Dr. Jaeger, under the name of *phytosaurus*. Two genera have recently been found near Bristol, by Dr. Riley and 'Mr. Stutchbury.

26. IMPRESSIONS OF THE FEET OF BIRDS AND QUADRUPEDS.—A few years since, the attention of geologists was called to the remarkable occurrence of the impressions of the feet of some unknown animal, in blocks of the New red sandstone, at Corn-cockle Muir, in Dumfriesshire; these prints appear to have been made by the paws of turtles, as I shall hereafter notice.† A discovery of a similar nature has since taken place in Saxony, in a quartzose

* Professor Owen, on the Anatomy of the Brachiopoda, Zoological Transactions, vol. i. p. 145.

† Account of the marks of footsteps of animals found impressed in sandstone, by the Rev. H. Duncan, D.D. Edinburgh. Trans. Royal Soc. vol. xi. 1828.

sandstone, at a village near Hildburghausen. These footsteps are of various sizes, and follow each other, being disposed in tracks, as if the animals had leisurely walked over the soft sand. Slabs of the stone, with their impressions, may be seen in the British Museum. They are supposed to be the foot-marks of some marsupial animal, like the kangaroo; but this is mere conjecture, and until bones shall have been discovered in the strata, but little reliance can be placed on the hypothesis, however ingenious. The largest marks are eight inches long, and five wide. No sooner had the savans of Europe been made acquainted with these phenomena, than a great number of relics, of a corresponding kind, were discovered in the United States, by Professor Hitchcock, of which an interesting account was published in the American Journal of Science, for 1836. In the laminated New red sandstone, which is spread over the valley of the Connecticut, numerous foot-prints appear on the surface of the sandstones when split asunder, exactly in the same manner as the ripples on the Tilgate grit (page 328). The impressions are supposed by Professor Hitchcock to be those of birds; and if this opinion be correct, it is clear that the feathered tribes of that ancient epoch were the iguanodons of their race, for the dimensions of one kind of foot-mark are *fifteen inches long*, exclusively of the large claws, which are two inches in length; a proportion twice as large as those of the ostrich.

The impressions are referrible to six or seven species, which differ in size and proportion.* No bones have been found in the stone; but scales of the same genus of fish (*Palæothrissum*) occur as in the saliferous sandstones of Europe. Until bones shall be discovered, the inferences as to the real nature of these impressions must be considered problematical; the fact, nevertheless, is very extraordinary, and will unquestionably lead to highly interesting results.

27. REPTILES.—I have reserved for this lecture some observations on the organization of reptiles, which may be necessary to enable the unscientific inquirer fully to comprehend the inferences that arise from an investigation of the fossil remains of this class of beings.

All animals possess organs by which a certain change is effected, by atmospheric agency, in the circulating fluid, to refit it for the purposes of nutrition. Land animals are furnished with an apparatus of cellular tissue, termed lungs, by which a large surface of the blood is brought in contact with the air; in aquatic animals, this apparatus forms gills, which are instruments fringed with innumerable processes, supplied by myriads of vessels, disposed like net-work, by which the blood is exposed to the action of aerated water, oxygen

* American Journal of Science, 1836.   Dr. Buckland's Essay contains an interesting account of these discoveries, illustrated by several plates.

absorbed, and the process of vitality maintained. In
reptiles, the respiratory organs are less developed
than in any of the other vertebrated animals; the
heart is so disposed, that at each contraction only a
portion of the volume of blood is sent to the lungs;
hence the action of oxygen on the circulating fluid
is in a less degree than in any of the mammalia,
birds, or fishes. As animal heat, the susceptibility
of the muscles to nervous influence, and even the
nature of the skin, are dependent on respiration, the
temperature of reptiles is low, and their muscular
powers are, on the whole, very inferior to those of
birds or mammalia; requiring no integuments, as
hair, wool, or feathers, to preserve their temperature, they are merely covered with scales, or have
a naked skin. As they can suspend respiration
without arresting the course of the blood, they
dive with facility, and remain under water for a long
period without inconvenience. They are oviparous,
laying their eggs, which they never hatch, on the
sands or banks. They present great diversity of
forms; some are extremely elegant, others grotesque
and hideous, and many have dermal processes of the
most fantastic shapes. Their habits are exceedingly variable; some are agile, others torpid; all
hybernate, or rather relapse periodically into a state
of dormancy, whether produced by cold, drought,
or excessive moisture. Their peculiar structure
enables them to endure long abstinence, to an extent
impossible to other races of animals. Their seasonal

habits, or, in other words, alternate periods of activity and repose, are in accordance with the sudden
evolution of the seasons in warm climates; they are
dormant when nature does not need their agency,
and rouse into activity when required to repress the
redundancy of those vegetables or animals which
constitute their food; exhibiting an admirable adaptation to the peculiar condition of existence which
they are destined to fulfil.   Some are herbivorous,
others carnivorous, and many prey on insects; their
powers of progression are as various—some orders,
though destitute of fins, wings, or feet, bound along
the ground with great agility; others walk or
swim; while some species are capable of flight.
Von Meyer, an eminent German naturalist, has
therefore arranged the fossil reptiles into groups,
characterised by the organs of motion;* as *flyers*,
swimmers, or *walkers*.   From this general view of
the economy and habits of the recent reptiles, we
shall be able to comprehend the physical conditions
required by those extinct forms which occur in a
fossil state, and thus arrive at some interesting conclusions respecting the regions which they inhabited.

28. TURTLES.—In turtles the want of active
faculties is compensated by their passive means of
resistance.   They have no weapons of offence, but

* Palæologica zur Geschichte der Erde und ihrer Geschöpfe
Von Meyer.   See a translation of a portion of this work, "On
the Structure of the Fossil Saurians," by G. F. Richardson.
Magazine of Nat. Hist. vol. i. p. 281.

are inclosed in a panoply of armour formed by the expansion of the ribs above, and by the bones of the chest beneath ; the carapace, or buckler, constituting the shell that spreads over the back of the turtle, is composed of the ribs, which, instead of being separated by intervals as in other animals, are spread out and united together.   Thus in the delicate and agile form of the serpent, and in the heavy and torpid mass of the turtle, the same gene-ral principles of structure prevail, and by a simple modification the skeleton is adapted for beings of very dissimilar forms and habits.   The *testudinata,* or turtles, like the other large reptiles, are essen-tially confined to torrid and temperate regions. The fresh-water species appear capable of bearing a higher latitude than the terrestrial : upon the whole the utmost range of this class of reptiles appears to be from 54° N. lat. to about 40° S. lat.*   The fluviatile species of tortoise, or *Emys,* are carni-vorous, feeding on frogs and small animals ; those of the genus *Trionyx, (three claws,)* are African or Asiatic, with the exception of the *Trionyx ferox,* which inhabits the hot regions of America. They live upon food which is found stationary at the bottoms of rivers ; in the stomachs of several procured from the Ganges, Col. Sykes found large quantities of muscles, the shells of which were broken into small

* On the Testudinata, by Thomas Bell, Esq. 1 vol. folio ; one of the most splendid works on Natural History that has appeared in this country.

angular fragments. I have fossil bones of a trionyx
(*T. Bakewelli*) from Tilgate Forest, imbedded in
a mass of shells of the same genus. The form of
the ribs, and other parts of the skeleton, differ in
the land, river, and marine genera, so that the
fossil remains can, for the most part, be readily
determined.

29. FOSSIL TURTLES.—The remains of turtles
are among the earliest indications of the reptile
tribe, and occur in the New red sandstone. The
impressions of their feet, or pats, are seen in the
quarries of that rock, in Dumfriesshire; and entire
tracks of these prints are preserved on the surface
of the stone; "one slab exhibits twenty-four
continuous impressions of feet, forming a regular
track, with six distinct repetitions of the marks of
each foot, the fore-foot being different from the
hind-foot: the appearance of five claws is discernible
in each fore-paw."* The remarks on ripple-marks
in the Wealden sandstone (page 328), and on the
impressions of human feet in stone (page 66), in
the previous lecture, render it unnecessary to explain this phenomenon. In the Lias and Oolite,
remains of this family have been found; the Wealden
contains marine, land, and freshwater species; and
the Chalk several marine turtles; bones of a trionyx have been found in the Kentish chalk, by
Mr. W. H. Bensted. In the tertiary deposites,

* Dr. Duncan.

both marine and fresh-water turtles occur : they are associated with the sivatherium in the Sub-Himalayas — with the mastodon in the Burmese Empire—with palæotheria in France—and with fruits and tropical plants in Sheppey ; their bones and eggs are daily becoming imbedded in the recent conglomerate of the Isle of Ascension.

30. CROCODILES.—This family affords the only living reptiles that approach in magnitude to the colossal forms of the fossil kingdom. The Egyptian crocodile, or alligator, is known to attain a large size; and the rivers of India are inhabited by gavials of enormous bulk, sometimes nearly thirty feet in length. The gavials are distinguished by the great length and slenderness of the beak, or muzzle. The nature of the teeth of these animals, and their mode of increase and renovation, have already been described (page 352). The vertebræ, or bones of the back, are convex behind, and concave in front, fitting into each other, like a ball and socket; a construction rarely found in the fossil species. All the vertebræ of crocodiles from the Wealden, are either flat or slightly concave at both extremities; and this character prevails, not only in the fossil animals of this family, but also in the lizards. A crocodile, nearly twenty feet long, has been discovered in the lias of Yorkshire ; teeth of crocodiles have been found in the Stonesfield slate ; two species of gavial in the Kimmeridge clay of Havre, by M. Alexandre Brong-niart ; and two species in the lias of Wirtemberg,

by Dr. Jaeger. Species of several related genera occur in the oolite and Jura limestone. The Wealden affords two or three species ; the chalk an equal number ; the London basin, a species with concavo-convex vertebræ ; and similar remains abound in the newer pliocene deposites of North and South America, the Sub-Himalayas, &c. Neither the Iguanodon, Megalosaurus, Hylæosaurus, Ichthyosaurus, or Plesiosaurus, have ball and socket vertebræ. This peculiarity of the vertebral column is very remarkable, and seems to indicate some general condition of the earth during the secondary period requiring such a modification of structure in the reptile tribes.* This family of reptiles, like the testudinata, extends through the vast periods of the secondary and tertiary eras to the present time. As the crocodiles frequent fresh-water, and not the sea, their remains testify the existence of regions watered by streams and rivers.

31. ICHTHYOSAURUS, (*fish-like lizard.*) In the lias of the west of England, bones and teeth, supposed to belong to crocodiles, had for many years excited attention ; but until 1814, when a considerable collection, from Dorsetshire, formed by Miss Mary Anning, was exhibited in London, no accurate investigation of these interesting relics had been attempted. Subsequently a great number of bones and skeletons have been found, numerous

* Appendix, K.

memoirs published, and the form and structure of the
original thoroughly investigated, by Mr. König, Sir
E. Home, Cuvier, Conybeare, De la Beche, and Dr.
Buckland.   Many   extraordinary   specimens   are
figured and described in the splendid work of Mr.
Hawkins, whose unrivalled collection of these remains is deposited in the British Museum.*   The
bones and skeletons so abundant in the lias are
chiefly referrible to two genera; the one called the
ichthyosaurus, by Mr. König, to denote its relation
to fishes and reptiles; the other, *plesiosaurus*, so
named by Mr. Conybeare, to mark a nearer approach to the lizard, or saurians, than the animals
of the other genus.

TAB. 54.—RESTORED FIGURE OF THE ICHTHYOSAURUS.

(*From Mr. Hawkins.*)

The ichthyosaurus had the beak of a porpoise,
the teeth of a crocodile, the head and sternum of a
lizard, the paddles of cetacea, and the vertebræ of
fish.   This restoration shows its general configura-
tion.   There are eight or more species, some of

* Memoirs of Ichthyosauri and Plesiosauri, extinct Mon-
sters of the ancient Earth; by Thomas Hawkins, Esq. F.G.S.
Folio, with 28 Plates.   Relfe and Fletcher, London.

which attain a magnitude equal to that of young whales. The teeth are conical, sharp, and striated, resembling those of crocodiles in the power of reproduction, but differing in the number, situation, and mode of regeneration; one species has 110 in the upper and 100 in the lower jaw. The orbit is very large, and the sclerotic, or outer coat of the eye made up of thin bony plates, arranged round the central opening or pupil, as in the owl and other birds; a mechanism by which the power of the eye is materially increased, and vision adapted to near or remote objects at will.* The bones forming the sternum or chest, which protect the organs of respiration, are very strong and largely developed, and those of the sternal arch offer a remarkable correspondence with those of the sternum of the Platypus. Like turtles, the animal had four paddles, composed of numerous bones enveloped in one fold of integument, so as to appear an entire fin, as in the cetacea; the forepaddles are large, and in some species are formed of one hundred bones; the hind are smaller, and contain but thirty or forty. The internal structure of these instruments, therefore, resembles that of turtles; and, as is even the case with the fin of the porpoise, the same elements of an arm are found as in the mammalia—a *humerus, radius, ulna,* and *phalanges.* The nostrils, which in crocodiles are situated at the extremity of the beak or muzzle,

* Dr. Buckland.

F F

are placed, as in the cetacea, beneath the orbits.
The vertebræ are hour-glass shaped, like those of the
sharks and other fishes; the spinal column, therefore,
admitted of the utmost freedom of motion; while
in the neck, the vertebræ connecting the head to
the spinal column are anchylosed, and have supple-
mentary bones to increase the strength and diminish
motion.*   The general figure of the ichthyosaurus
must have been that of a grampus or porpoise,
having four large fins or paddles.   The teeth prove
it to have been carnivorous; the paddles, that it
was aquatic; the scales, bones, and other remains,
constantly found in the interior of the skeleton, that
it was an inhabitant of the sea.   Its skin appears
not to have been covered with scales.†

TAB. 55.—THE PLESIOSAURUS.

(*From Mr. Hawkins.*)

32. THE PLESIOSAURUS.—The discovery of a
remarkable specimen, by Miss Anning, enabled
Mr. Conybeare to establish the character of that

* Memoir on a peculiarity of Structure in the Neck of the
Ichthyosaurus; by Sir P. M. de Grey Egerton, Bart.

† The *epidermis*, or scarf-skin, and the *corium*, or true skin,
occur in a fossil state.—Dr. Buckland, Plate 10, Fig. A. 1, 2,
3, 4.

extraordinary creature, the Plesiosaurus, which
differs from the ichthyosaurus in the extreme small-
ness of the head, and enormous length of the neck;
the latter is composed of upwards of thirty vertebræ,
a number far exceeding that of the cervical vertebræ
in any known animal. This reptile combines in its
structure the head of a lizard with teeth like those
of a crocodile, a neck resembling the body of a
serpent, a trunk and tail of the proportions of those
of a quadruped, and paddles like those of turtles.
It has been compared to a serpent threaded through
the body of a turtle. The vertebræ are longer and
less concave than those of the ichthyosaurus, and
the ribs, being connected by transverse abdominal
processes, present a close analogy to those of the
cameleon.

There are six or more known species of Plesiosauri. The collection of Mr. Hawkins, in the
British Museum, contains a skeleton eleven feet
long, and so nearly perfect, that the form of the
original creature may readily be traced. Mr. Conybeare compares the Plesiosaurus to a turtle stripped
of its shell, and thinks it probable, from its long
neck presenting considerable impediment to rapid
progress in the water, that it frequented the coast,
and lurked among the weeds in shallow water. As
it is evident that it must have required frequent
respiration, it probably swam on or near the surface, and darted down upon the fishes on which it
preyed.

Ichthyosauri and Plesiosauri have been found throughout the secondary strata, from the lias to the chalk inclusive; Lyme Regis, owing to the researches of Miss Anning, is the most celebrated locality in England, but they also occur in the lias of other parts of Dorsetshire, and of the adjacent counties. Their remains have been discovered in the oolite both of this country and of the continent; I have found vertebræ of the Plesiosaurus in the green sand of Berkshire, and in the Wealden. Dr. Harlan describes bones and teeth, which he refers to Ichthyosauri and Plesiosauri, from the secondary formations of the United States.*

33. PTERODACTYLES, or FLYING REPTILES.— Of all the wonderful beings which the researches into fossil osteology have brought to light, the Pterodactyles are unquestionably the most extraordinary. With a head and length of neck resem-

---

* Medical and Physical Researches; by B. Harlan, M.D. Philadelphia. In this interesting volume, in addition to notices of several fossil saurians allied to the crocodile, ichthyosaurus, &c. Dr. H. figures and describes bones and teeth of an enormous reptile, which he names *Basilosaurus*, and he has with great liberality presented me with one vertebra, which is nine inches in length. Neither the relation of this animal to other species, or its geological position, has been accurately determined, but Dr. Harlan expresses his expectation of additional and more illustrative specimens.

A list of the known species of Ichthyosauri and Plesiosauri, and of their localities, is given by Mr. de la Beche, Geological Manual, page 365.

bling those of a bird, the wings of a bat, and the body and tail of ordinary mammalia, these creatures present an anomaly of structure as unlike their fossil contemporaries, as is the duck-billed platypus, or ornithorhynchus of Australia, to existing animals. The skull is small, with very long beaks, which extend like those of a crocodile, and are furnished with upwards of sixty sharp-pointed teeth.; the size of the orbit denotes a large eye, and it is therefore probable that these creatures, like other insectivora, were nocturnal. The fore-finger is immensely elongated for the support of a membranous expansion, as in the bat: impressions of the web are seen in some specimens. The fingers terminated in long hooks, like the curved claws of the bat. The size and form of the foot, leg, and thigh, show that the Pterodactyle was capable of perching on trees, and of standing firmly on the ground, where, with its wings folded, it might walk or hop like birds.* This stuffed specimen of an enormous bat. from the museum of the late Mr. Brookes, will serve to convey some idea of the fossil animal ; I would here observe, that the wing of the bat is not merely an instrument for flight, but its structure is so exquisite, and the web so abundantly furnished with nerves, that the organ seems to possess a peculiar sensation, by which the bat, although moving with

* Dr. Buckland. Mr. Martin has introduced a restored figure of the Pterodactyle in the foreground of the Frontispiece of this volume.

the utmost rapidity, is enabled to avoid objects in its flight. Eight species of Pterodactyles have been discovered, and these vary in size from that of a snipe, to a cormorant. At Solenhofen the bones of Pterodactyles are associated with the remains of *libellulæ*, or dragon-flies;* and in the Stonesfield slate with the *elytra*, or wing-cases, of beetles. The remains of a species of the size of a raven, discovered in the lias of Lyme Regis by Miss Anning, to whose talents and indefatigable researches British Palæontology is so deeply indebted, have been described and admirably elucidated by Dr. Buckland:† this specimen is in the British Museum.

Numerous thin delicate bones, evidently belonging to Pterodactyles, have been found in the Stones-field strata; and in the Jura limestone, seven species, which have been determined by Professor Goldfuss. In the Wealden, bones of Pterodactyles occur, and afford proof that these extraordinary creatures inhabited the country of the Iguanodon.

34. FOSSIL SALAMANDER.—Fossil remains of toads, frogs, and salamanders, have been found in some of the newer tertiary strata. The quarries of Œningen that yielded the fossil fox, mentioned in a former lecture (page 233), have also afforded specimens of the batrachian family of reptiles. The most celebrated relic of this kind is one which a German physician of some note, Scheuchzer, in

* Goldfuss.   † Geological Transactions, vol. iii. page 220.

1726, declared to be a fossil man.* Cuvier, however, has shown that it is the skeleton of a gigantic,
aquatic salamander, of an unknown species ; a very
fine portion of the skeleton of the same species, from
Œningen, is in the British Museum.   It is worthy
of remark, as a proof how prejudice of opinion may
blind us to the most obvious truths, that Scheuchzer, a physician, and therefore conversant with
human osteology, could yet so grossly deceive
himself, as to believe that the fossil bones were
those of a man ; he describes the specimen, in an
essay entitled *Homo Diluvii testis*, as being indis-
putably the moiety, or nearly so, of a human
skeleton, and states that the bones, and even substance of the flesh, are incorporated in the stone ;
and that it is a relic of that "cursed race which
was overwhelmed by the deluge !"

35. FOSSIL REPTILES ALLIED TO THE LIZARDS.
—The Iguanodon, 100 feet in length, to which the
Iguana, but one-twentieth of the magnitude, is the
only recent species that bears any affinity ; and the
Megalosaurus, sixty feet long, whose only living
analogue, the Monitor, never attains a greater
length than six feet, have been cursorily noticed ;
and I am compelled to pass them by in this hasty
review without dwelling on many novel and highly
interesting peculiarities of structure which their
fossil remains present.   They were both terrestrial ;

---

* Philosophical Transactions, vol xxxiv.

the one was decidedly herbivorous; the other, from
its flat, curved, serrated, pointed teeth, there is
reason to conclude was carnivorous. In the magnesian conglomerate on Durdham Downs, near
Bristol, three distinct species of fossil Saurians,
related to the Iguana and Monitor, have lately
been discovered by Dr. Riley, and Mr. S. Stutchbury. But I cannot dwell on this or other notices
of fossil saurians; for so numerous have been the
recent discoveries of reptilian remains, that a bare
enumeration of the essays that have been published on the subject would encroach too far on
my limits.*

36. REVIEW OF THE AGE OF REPTILES.—From
the examination of the organic remains of the
secondary formations we arrive at the following
results:—that the seas, lakes, and rivers, during the
geological epoch termed secondary, were peopled
by fishes, mollusca, crustacea, radiaria, polyparia,
and other zoophytes; all of extinct species, and
presenting as a whole, a greater discrepancy with
existing forms, than those of the tertiary; the most
remarkable feature being the absence of cetacea,
and the presence of several genera of extinct marine
reptiles. On the land we find no analogy to the

* Consult Cuvier's Recherches sur les Ossemens Fossiles,
tom. v.;—Pidgeon's translation of the Fossil Animal Kingdom,
1 vol. 8vo. 1830;—and Dr. Buckland's Essay, which is a fund of
instruction of the highest interest, conveyed in the most en-
gaging style.

tertiary or present eras; throughout the vast accumulations of the spoils of the ancient islands and continents, although the remains of crocodiles, fresh-water turtles, insects, and terrestrial plants abound, one or two jaws of a small animal related to the opossum are the sole indications of the existence of mammalia; and the bones of a species of wader, the only evidence of the presence of birds. In vain we seek for the bones of man, or the remains of works of art—for the skeletons of the mastodon or of the elk—of the palæotheria, or of other mammalia that were their contemporaries; the osseous remains of terrestrial or fluviatile reptiles alone appear. Here then, in the language of Cuvier; "Nous remontons à un autre âge du monde —à cet âge où la terre n'était encore parcourue que par des reptiles à sang froid; où la mer abondait en ammonites, en bélemnites, en térébratules, en encrinites; et où tous ces genres, aujourd'hui d'une rareté prodigieuse, faisaient le fond de sa population."*

We have seen that in the carboniferous limestone, the lowermost or most ancient of the formations in which reptiles occur, turtles, and several genera related to the lizards, have been discovered—in the lias, swarms of extinct marine reptiles, the ichthyosauri and plesiosauri, with turtles, crocodiles, and pterodactyles—in the oolite, the megalosaurus, and

* Oss. Foss. tom. v.

several new genera allied to the crocodiles, and *one* genus of *mammalia*—in the Wealden, the iguanodon, hylæosaurus, and other related genera, and *one genus of birds*. Thus the fauna of the secondary epoch, as established by its organic remains, presents the following characters :—

Mammalia . One marsupial animal.

Birds . . . . { One species of wader.
{ Supposed impressions of the feet of several species.

Reptiles . . { Marine—about twelve genera, including thirty or more species.
{ Fluviatile and terrestrial—ten genera, with twenty species.
{ Flying—one genus, eight or nine species.

Insects . . . . Several species of libellula, and some coleoptera.

If we admit, to the utmost extent, the effect of causes that can be supposed to have operated in the exclusion of mammalian remains from the deposites under investigation, still the overwhelming preponderance of the reptile tribes, both on the land and in the waters, is most striking.   But does this remarkable phenomenon support the hypothesis which some geologists have advanced, that during the secondary epoch, the earth was not adapted to the existence of mammalia; that it was in a state of turbulence and convulsion, which colossal reptile forms were alone calculated to endure—that it was a half-finished planet, unsuitable to warm-blooded animals, and its atmosphere incapable of supporting the higher types of organization ?   The proof that birds existed in the country of the iguanodon—that marsupial animals inhabited the region of the megalosaurus and pterodactyle—that trees and plants, related to genera which now grow in territories abounding in mammalia, flourished in the dry land of that ancient epoch, are facts which appear to me fatal to such a hypothesis, and show that the physical condition of the earth, seas, and atmosphere, was not essentially different from that of the tertiary and modern periods.

That reptiles predominated throughout the secondary epoch, to a degree far beyond what has since prevailed, cannot, by any legitimate process of reasoning, be disputed; but I do not think we are yet in possession of data by which the problem can be solved.

37. OBJECTIONS ANSWERED.—There are some who, with one of the Bridgewater essayists (Mr. Kirby*), oppose these conclusions, and have recourse to the most strange conceits to account for the phenomena on which they are founded. But it is for those who refuse their assent to deductions made with the greatest caution, and derived from an overwhelming mass of evidence, to explain the entire absence of

---

* Seventh Bridgewater Essay. Mr. Kirby supposes there is a subterranean world of reptiles, where the Iguanodon still flourishes!!! and that the occurrence of a vertebra of the Ichthyosaurus in diluvial gravel is a proof of the modern existence of that reptile ! As Dr. Buckland's Essay follows that of Mr. Kirby, the reader has the bane and antidote both before him.

all traces, not only of man, but of the whole existing
species of animals and vegetables in the ancient
deposites; while there is not a river, or even stream,
which does not daily imbed the remains of the
present inhabitants of the globe. But however
future discoveries may modify this hypothesis, they
cannot invalidate the fact, that there is no country
on the face of the earth with such an assemblage of
animal life, as that possessed by the regions whence
the delta of the Wealden was derived; no where
is there an island or a continent inhabited by
colossal reptiles only, or where reptiles usurp the
place of the large mammalia. We have seen that
this feature in the zoology of that remote period was
not confined to the country of the Iguanodon; in
every part of the world where geological researches
have extended, this wonderful phenomenon appears—
the absence of mammiferous animals. The bones of
reptiles, of enormous size, are the only animal remains that occur in any considerable number. It
is, therefore, certain that there was a period when
oviparous quadrupeds, of appalling magnitude, were
the chief possessors of the lands, of which any
traces remain in the strata that are accessible to
human observation. I do not mean to assert,
that reptiles, and reptiles only, were the occupiers
of every island and continent, but I have shown, by
the most irrefragable testimony, that the reptile
tribes, during the secondary periods, were developed
to an extent of which the present state of animated

nature affords no example. I am ready to acknow-
ledge that the proposition is somewhat astounding,
and I do not feel much surprise that many intelligent persons hesitate to admit its correctness; but
you have seen that it is deduced from such an
immense accumulation of facts, as to compel assent,
in spite of all our preconceived opinions. We may,
indeed, call up from the depths of our ignorance
hypotheses as marvellous as the phenomena they
are intended to explain, but which a very slight
examination of the facts described would prove to
be utterly untenable.

38. CONCLUDING REMARKS. — There is another objection to which I would allude, for I
do not think with some, that the errors, or even
prejudices, of those who differ from us should be
treated with silence or contempt; but, rather,
that it is our duty to explain, again and again, the
foundation of our belief, in the hope and assurance
that we shall at length either remove the erroneous
opinions of others, or be convinced of the fallacy of
our own. It has been insisted upon by those whose
views are limited to the present state of the globe,
that the supposition of the earth having been peopled
by other creatures before the existence of man, is
incompatible with the evident design of the Creator,
and derogatory from the dignity of the human race,
to whose pleasures and necessities all things were
rendered subservient, and for whom alone they
were created. But this assumption is utterly at

variance with what we know of the living world
around us; every where we see forms of animated
existence, possessing faculties and sensations wholly
dissimilar to our own ; and while, in the beautiful
language of Scripture, we are told that not a sparrow
falls to the ground without our heavenly Father's
notice, the contemplation of the present constitution
of nature, by any philosophical observer, would
alike condemn such vanity and presumption. For
my own part, feeling, as I do, the most profound
reverence, and the deepest gratitude to the Eternal,
who has given unto me this reasoning intellect,
however feeble it may be ; and believing that the
gratification and delight experienced in the contemplation of the wonders of creation here, are but a
foretaste of the inexpressible felicity which, in a
higher state of existence, may be our portion hereafter ; I cannot but think that the minutest living
atom, which the aided eye of man is able to explore,
is designed for its own peculiar sphere of enjoyment,
and is alike the object of His mercy and His care,
as the most stupendous and exalted of His creatures.
In nothing, perhaps, are we more mistaken, than
in our estimate of the happiness enjoyed by other
beings ; to employ the beautiful simile of a distinguished writer*—" As the moon plays upon the
waves, and seems to our eyes to favour with a peculiar beam one long track amidst the waters, leaving

* Bulwer.

the rest in comparative obscurity, yet all the while she is no niggard in her lustre—for although the rays that meet not our eyes, seem to us as though they were not, yet she, with an equal and unfavouring loveliness, mirrors herself on every wave; even so, perhaps, happiness falls with the same power and brightness over the whole expanse of being, although to our limited perception it seems only to rest on those billows from which the rays are reflected back upon our sight." And if we admit, as all must admit who for one moment consider the marvels which astronomy has unfolded to us, that there are countless worlds around us, inhabited by intelligences, of whose nature we can form no just conception, surely, the discoveries of geology ought not to be rejected because they instruct us that ere man was called into existence, this planet was the object of the Almighty's care, and teeming with life and happiness.

This, then, is the sublime truth revealed to us by geology—*that for countless ages our globe was the abode of myriads of living forms of happiness, enjoying all the blessings of existence, and at the same time accumulating materials which should render the earth, in after ages, a fit, temporary abode, for an intellectual and immortal being!*

LECTURE VI.

1. INTRODUCTION. — The secondary formations reviewed in the last discourse, presented a marked increase in those extraordinary types of animal life —the zoophytes, polyparia, and crinoidea ; some beds, as the coral-rag of the oolite, consisted almost entirely of corals ; while in other deposites the mineralized skeletons of the lily-shaped radiaria

were not less abundant. As we advance to the more ancient rocks, we shall find that the remains of these animals prevailed in the older secondary formations, almost to the exclusion of other families, and that entire ranges of mountains are composed of the consolidated *debris* of corals; and vast beds of limestone and marble, of the petrified skeletons of crinoidea. That we may understand the nature of these deposites, and be enabled to arrive at accurate conclusions as to their formation, a knowledge of the structure and habits of the recent animals is necessary, and I therefore purpose devoting the present discourse to a general and familiar exposition of the natural history of the recent and fossil corals, and of the lily-shaped animals.

2. ORGANIC AND INORGANIC KINGDOMS.—The beautiful world in which we are placed, is every where full of objects presenting innumerable varieties of form and structure, of action and position ; some of them being inanimate or inorganic, and others possessing organization or vitality. The organic kingdom of nature, in like manner, is separated into two grand divisions, the animal and vegetable. The differences between organic and inorganic bodies are numerous and manifest ; but it will suffice for my present purpose to mention a few obvious and familiar characters. All the parts of an inorganic body enjoy an independent existence ; if I break off a crystal from this mass, the specimen does not lose any of its properties, it is

G G

still a mass of crystals as before; but if a branch be
removed from a tree, or a limb from an animal,
both are rendered imperfect, and the parts removed
suffer decomposition,—the branch withers, and the
animal matter undergoes putrefaction. But if crystals, which may be considered the most perfect
models of inorganic substances, be formed, they
will continue the same, unless acted upon by some
external force of a chemical or mechanical nature.
Within, every particle is at rest, nor do they possess
the power to alter, increase, or diminish : they can
augment by external additions only, and decrease
but by the removal of portions of their mass.*   But

    * These remarks must be taken in a general sense only, since
recent experiments have demonstrated that the molecules of
inorganic matter undergo modification by the slightest change
of temperature.

    " Prismatic crystals of zinc are changed in a few seconds into
octahedrons by the heat of the sun. We are led from the mo-
bility of fluids to expect great changes in the relative positions
of their molecules, which must be in perpetual motion even in
the stillest water or calmest air; but we were not prepared to
find motion to such an extent in the interior of solids. We
knew that their particles were brought nearer by cold and pressure, or removed farther from one another by heat; but it could
not have been anticipated that their relative positions could be
so entirely changed as to alter their mode of aggregation. It
follows from the low temperature at which these changes are
effected, that there is probably no portion of inorganic matter
that is not in a state of relative motion. Prismatic crystals of
sulphate of nickel exposed to the summer heat, in a close vessel,
had their internal structure completely altered, so that when
broken open they were composed internally of octahedrons, with

organic bodies have characters of a totally different nature; they possess definite forms and structures, which are capable of resisting for a time the ordinary laws by which the changes of inorganic matter are regulated, while internally they are in constant mutation. From the first moment of the existence of the plant or animal to the period of its dissolution there is no repose; youth follows infancy,—maturity precedes age; it is thus with the moss and the oak, — the monad and the elephant, — life and death are common to them all. Animals and vegetables also require a supply of food and air, and a suitable temperature, for the continuance of their existence; and they are nourished by particles prepared in appropriate organs, and conveyed by suitable vessels. From the very first germ of an animal or a vegetable, there is a vital principle in action, by which are developed in succession the ordained phenomena of its existence. By this power the germ is able to attract towards it particles of inanimate matter, and bestow on them an arrangement widely different from that which the laws of chemistry or mechanics could produce. The same power not only attracts these particles, and pre-serves them in their new situations, but is continually engaged in removing those which might by their presence prevent or derange its operations;

square bases. The original aggregation of the internal particles had been dissolved, and a disposition given to arrange themselves in a crystalline form."—*Mrs. Somerville*, p. 171.

and, on the other hand, so soon as the vital principle deserts the body which it has animated, the
latter immediately becomes subject to the agencies
which act on inorganic matter: " in obedience to
the power of gravitation the bough hangs down,
and the slender stem bends towards the earth,—the
animal falls to the ground,—the pressure of the
upper parts flattens those on which they rest,—
the skin becomes distended, and the graceful outlines of life are changed for the oblateness of
death," * — the laws of chemistry then begin to
operate, — putrefaction takes place, — and, finally,
dust returns to dust, and the spirit of man to Him
who gave it.

3. DIFFERENCE BETWEEN ANIMALS AND VEGE-
TABLES.—I have thus briefly described a few of the
phenomena peculiar to organic existence; it will
now be necessary to offer some remarks on the
distinguishing characters of the animal and vegetable kingdoms, for unless we have a clear perception of the phenomena peculiar to each, we shall not
obtain correct ideas of the nature of zoophytal
organization.

When we compare together those animals and
vegetables which are considered as occupying the
highest stations in each kingdom, we perceive that
they differ from each other in particulars so obvious
and striking, as not to admit of question. The horse,

* Dr. Fleming.   Philosophy of Zoology, 2 vols. 8vo.

and the grass upon which it feeds; the bird, and the tree in which it builds its nest, are so essentially distinct from each other, that we perceive at once that they belong to distinct classes of organic nature. But it is far otherwise when we descend to those animals and plants which occupy the lowest stations in vitality: here the functions to be performed are but few, the points of difference obscure, and it requires a correct knowledge of the laws of organization, and a careful application of that knowledge, to enable us to determine with precision where animal life terminates, and vegetable existence begins. The lichen which grows on the stone, and the flustra attached to the rock, present but little difference to the common observer; both are permanently fixed to the spot on which they grow, from the earliest period of their existence to their dissolution; and in the vegetable dried by the heat of the sun, and in the coralline shrivelled up from the absence of moisture during the ebb of the tide, we might seek in vain for those characters, which would assign the one to the vegetable, and the other to the animal kingdom.

4. NERVOUS SYSTEM AND SENSATION.—My limits will not permit me to dwell on the obvious distinctions which exist between animals and vegetables in their chemical composition, and in the form and distribution of their vessels. I must content myself with mentioning the more important character which animals alone possess—the faculty

of SENSATION, communicated to animal matter by
a nervous system. In vertebrated animals a brain
and spinal marrow form the apparatus by which
nervous influence is developed.

Thus when any object comes in contact with my
finger I am sensible of its presence, and my finger is
said to possess sensation ; if I compress or cut across
the nerve which passes from the brain to the finger,
this faculty of sensation is suspended or destroyed :
the same object may come in contact with my
finger as before, but no feeling is excited indicating
to me its presence. This phenomenon must be
familiar, for every one must, in lying or sitting,
have compressed the nerve of the arm or thigh, and
occasioned a temporary numbness and loss of accurate feeling in the limb. I perceive, then, by my
own experience, that the power of feeling is insepa-
rably connected with the presence and condition of
the nerves ; and that in man, and the higher classes
of animals, this nervous influence is transmitted
from the brain and spinal marrow.

In examining the other divisions of the animal
kingdom, the presence of a nervous system, more
or less developed, may be detected : in the animals
of the higher orders, nervous filaments can be distinctly traced, from their origin to their distribution
in the various parts to which they communicate
sensation. But in proportion as the system of ab-
sorbing, secreting, and circulating vessels, becomes
less, a corresponding diminution takes place in the

nervous fibres, till at length both the vessels and nervous filaments elude our finite observation, and we are left to infer from analogy, that, since sensation depends on the presence of nerves, and the smallest animals evidently possess sensation, a nervous system exists in the minutest monad of animal organization.

In the largest and most perfect examples of the vegetable kingdom, no traces of nerves are perceptible, nor of any substance which can be considered as at all analogous in structure or function : it is therefore concluded, that as vegetables are destitute of nerves, they are likewise wanting in that faculty which in animals we term sensation.

But the nerves not only bestow feeling, they also confer the power of voluntary motion; and, if the construction of the organs to which such nerves proceed be suitable, they enable the animal to effect progression, or in other words, give it the faculty of changing its situation from one place to another. As we descend in the scale of creation, we find many animals destitute of that power, and living on the same spot from the commencement to the termination of their existence; and all these animals are inhabitants of the waters.

Such, then, are the essential characters of animal existence—an external determinate form, gradually developed, with an internal organization possessing circulating vessels for effecting nutrition and support, and capable of attracting and assimilating

particles of inorganic matter, combined with a nervous system communicating sensation and voluntary
motion; a certain term of existence being assigned
to determinate forms—in other words, a period of
life and death.

5. DIVERSITY OF ANIMAL FORMS.—The form
is as varied in figure and magnitude as the imagination can conceive; from the god-like image of man
to the shapeless mass of jelly that floats upon the
wave—from the elephant and the whale to the
insect and the animalcule, of which five hundred
millions may be contained in a drop of water. In
fact, so numerous and dissimilar are the modes of
animal existence on the globe, that the opinion of
astronomers that the inhabitants of the worlds
around us, must, from the different densities and
conditions of the respective planets, be totally distinct and unlike any that exist on the earth, ought
not to be considered marvellous or incredible.
But of all the shapes in which animal existence
presents itself, none are more extraordinary, or
unlike what is commonly conceived of living beings,
than those compound creatures which are described
by naturalists under the name of *zoophytes*, or
animal plants, and familiarly known in their varied
forms by the names of corals, madrepores, sponges,
sea-anemones, dead men's fingers, &c.

6. ELLIS'S DISCOVERIES.—It was in this town
(Brighton), in the year 1752, that Ellis first ascertained the animal nature of several of the small

corallines and sponges which abound on our shores.
It appears that he was engaged in forming a collection of marine plants for the instruction of the
young princesses in botany, and having occasion to
examine some of the specimens through a powerful
microscope, he was astonished to find that the
sponges, which were then supposed to be marine
plants, possessed a system of vessels through which
the sea-water circulated; and that many of the
corallines exhibited pores, from which tentacula or
feelers were constantly protruding, and then suddenly retracting, as if seizing and devouring prey.
Subsequent observations have confirmed his opinion,
that sponge is an animal; and that the substance
we call by that name is the skeleton, or support, of
a vascular substance which invests it, and which
may be considered as the body of the animal.
When viewed through the microscope, innumerable
pores are seen on the surface constantly imbibing
salt water, which circulates throughout the mass,
and is finally rejected from the large pores; this
water doubtless contains the living atoms which
constitute the food of this compound animal, but
are so minute as to elude our observation.

7. SPONGE.—This simple form of animal existence
approaches so nearly to that of plants, that it will be
instructive to dwell a few moments on the investigation of its structure. The live sponge, when
viewed through a microscope, exhibits a cellular
tissue, permeated by innumerable pores, which unite

into cells, or tubes, that ramify through the mass
in every direction, and terminate in larger openings.
The minute pores through which the water is im-
bibed, have a fine transverse gelatinous net-work,
and projecting spicula, by which large animalcules
or noxious particles are excluded. Water incessantly enters into the smaller pores, traverses the
longer tubes, and is finally ejected from the larger
openings. This enlarged drawing, from Dr. Grant
(Pl. 1, fig. 10), of a section of a sessile sponge,
common on our shores, explains the mechanism of
these animals. But the pores of the sponge have
not the power of contracting and expanding, as
Ellis supposed; the water is attracted to these
openings by the action of instruments of a most
extraordinary nature, by which currents are produced in the fluid, and propelled in the direction
required by the economy of the animal.

8. CILIA, or VIBRATILE ORGANS.*—Although
these organs, which are termed *cilia*, or hair-like
instruments, are not confined to the class of animals
which form the subject of this inquiry, yet, as they
play so important a part in the economy of the
zoophytes and crinoidea, it will be necessary to
define their structure and functions; and I shall
avail myself of the highly interesting remarks of
my friend Dr. Grant, and of Dr. Sharpley, on this
subject, as well as on the anatomy and physiology

* From *Cilium*—eye-lash.

of the polyparia, hereafter to be noticed.* The cilia resemble very minute hairs, and are only visible with the microscope; they are situated in parts habitually in contact with water or other fluids, and possess the power of vibrating with great celerity, by which they produce motion and currents in the surrounding fluid. When a drop of water containing infusoria is brought under the microscope, it is seen that as these animals move along, every particle of foreign matter near them is agitated, a phenomenon indicating eddies in the water. When the animals remain stationary, the currents are more distinct, and evidently take certain directions, and cause the particles of matter to run in a stream to and from the animal. If a very high magnifying power be employed, transparent filaments will be distinguished projecting from the surface of the animalcules, and moving with extreme rapidity. These are the cilia, which serve as fins to assist the animal in progression; and when it is stationary, impel the water in currents through the cavities and tubes on which they are distributed : these must not be mistaken for the tentacula or feelers,

* On the Nervous System of the Beroë pileus, Zoological Transactions, vol. i. p. 10.   Outlines of Comparative Anatomy, by Robert E. Grant, M.D. F.R.S. Professor of Comparative Anatomy at the London University.

See the article *Cilia*, by Dr. Sharpley, Cyclopædia of Anatomy and Physiology.

but may be considered as fringes of delicate hair, investing those instruments, and the internal surfaces of other organs.   The cilia are so minute, that their outward form, position, and the direction of their motions, only can be detected, their internal structure eluding observation.   In the simpler forms of animals, the cilia are the organs for motion, respiration, and the obtaining of food.   They move with great regularity and velocity, and are exceed-ingly numerous ; Dr. Grant has calculated four hundred millions of them on a single flustra foliacea ! and has detected their existence in the sponge, and shown that the currents which incessantly flow into the pores of that animal, are produced by the vibrations of the cilia attached to the inner surface of the tubes.

9. FLUSTRA.—If we extend our observations to the patches of white calcareous matter, called *flustræ*, that may be seen on every sea-weed or shell on the shore (Pl. 1, fig. 12), appearing like delicate lace, we shall discover that these apparently mere specks of earthy substance, also belong to the animal kingdom.   Many species of flustræ are common along our coast, and I will describe their structure somewhat in detail, as their examination will serve to illustrate the nature of those species which, from their magnitude and immense extent, become such important agents in the economy of nature.

The flustra, when taken fresh and alive out of the water, presents to the naked eye the appearance

of fine net-work, coated over with a glossy varnish.
With a glass of moderate powers, it is discovered to
be full of pores, disposed with much regularity.
If a powerful lens be employed, while the flustra
is immersed in sea-water, very different phenomena
appear ; the surface is seen to be invested with a fleshy
or gelatinous substance, and every pore to be the
opening of a cell or cavity, whence issues a tube,
with several long feelers or tentacula : these expand,
then suddenly close, withdraw into the cells, and
again issue forth ; and the whole surface of the
flustra is studded with these hydra-like forms,
sporting about in all the activity and energy of
life.

The surface of the flustra, as seen under a moderate power (Pl. 1, fig. 7), exhibits a series of cells,
arranged in a regular manner, their forms and dis-
positions varying in different species. When highly
magnified, each cavity is found to be the receptacle
of a polypus*, which appears to be a transparent
jelly-like mass, having a receptacle or sac, the
external opening of which is surrounded by eight
or ten feelers or tentacula, that have the power of
extending and retracting with great rapidity. A
still higher power discovers that these tentacula are
in many zoophytes furnished with the vibratory
organs above described, and the existence of similar

* Polype, or polypus (many-feet), is a name derived from
the tentacula, or processes which in some species serve for progression, in others for respiration.

instruments is inferred in the minute species where the cilia have not yet been detected, because these atoms present the same phenomena of currents as the larger polyparia. The appearance of the polypi, in various states of expansion and contraction, is shown in these sketches (Pl. 1), after drawings made by Mr. Lister, from observations on living specimens, carried on during a short residence in this town.* The animalcules were kept alive by an ingenious contrivance, that secured a constant supply of fresh sea-water, without which they speedily perished, probably from want of food. The microscope, which was of unrivalled excellence, was so adapted that the polypi were kept constantly within the field of view, and I had the pleasure of witnessing the phenomena I am about to describe. Sometimes the polypus was seen protruding its tentacula from the cell, as in Pl. 1, fig. 2 ; in other examples, the animals retreated into their cells, and might be seen in different states of contraction (Pl. 2, fig. 9). In a specimen of *Campanularia* (Pl. 1, fig. 4), an animalcule is shown expanded in one branch, and re-tracted within the cell in the other. In the former instance, the tentacula were spread out, and nu-merous currents of water, evidently produced by the vibrations of the cilia, were seen rushing to-

* The Philosophical Transactions for 1834 contain a memoir on the structure and functions of the tubular and cellular polypi, by J. L. Lister, Esq. F.R.S., in which are detailed the results of his observations made at Brighton.

wards the centre. By these eddies the invisible beings which constitute the food of the flustra were brought within the vortex; the tentacula then closed, seized their prey, and conveyed it to the sac or stomach.

10. THE FOOD OF INFUSORIA.—However improbable it may appear to the mind unaccustomed to investigate the works of the Creator, that beings so minute as those under examination should prey upon living forms, of yet more infinitesimal proportions, the fact is nevertheless unquestionable. It is even possible to select the food of animacules much smaller than the polypi of the flustra, and thus exhibit their internal structure ! The animals called monads may be considered as the lowest limit of animated nature, so far as cognizable by man, their diameters varying from the twelve hundredth part of an inch to the *twenty-four thousandth ;* and the powers of the microscope will extend no farther. These creatures are of a cylindrical or spherical form, having a mouth by which their nutriment is taken in, and a stomach or digestive apparatus: the latter is visible only when these living atoms are fed with colouring particles, the animals being transparent and colourless, and their natural food equally so. Dr. Ehrenberg, of Berlin, by furnishing these infusoria with colouring matter for nourishment, has been able to illustrate their organization in a most extraordinary degree. He employed a solution of pure indigo for this purpose ; and the results of his

experiments are highly interesting. Immediately on a minute particle of a very attenuated solution of indigo being applied to a drop of water containing some of the pedunculated *vorticellæ*,* the most beautiful phenomena are observable. Currents are excited in the fluid in all directions, by the rapid motion of the cilia, which form a crown round the anterior part of the body of the animalcule, and the particles of indigo are seen moving in different directions, but generally all converging towards the orifice or mouth, situated not in the centre of the crown of cilia, but between the two rows of these organs, which exist consecutive to one another. The attention is no sooner drawn to this beautiful phenomenon than presently the body of the animal, which had been quite transparent, becomes dotted with a number of distinctly circumscribed spots, of a dark blue colour, exactly corresponding to that of the moving particles of indigo. In some species, particularly in those which are provided with an annular contraction or neck, separating the head from the body, the molecules of indigo can be

* *Vorticellæ.* The rotifera, or wheel-polypi, as they are commonly termed, from the supposition that they have organs which move round like a wheel, have cilia disposed in circles, which seen in some directions, when moving with great velocity, appear like wheels.—*Encyclop. Anat. and Phys.* p. 607. A highly interesting and lucid account of these animals, and of the whole family of infusoria and polyparia, is given by Dr. Roget, in his Bridgewater Essay ; a work of great labour and profound research.

traced in a continuous line in their progress from the mouth to the internal cavities.*

But although, as I have above remarked, the monad, in the present state of our knowledge, and with the wonderful instruments which the ingenuity of man has constructed, is the lowest term of organization, yet it is impossible to doubt that there are countless living forms concealed from our observation, some of which serve as food to these "miniatures of life." I may also observe that the structure of many of the animalcules is as varied and complicated as in the larger, and, to our imperfect conceptions, more important orders of animals.†

11. MODE OF INCREASE AND DEATH.—To return to the flustra :—If our observations on the living polypi be continued for a sufficient period, we shall at length perceive a small globule thrown off from the mass, and become attached to the

* " The size of the ultimate particles of matter must be small in the extreme. Organized beings, possessing life and all its functions, have been discovered so small that a million of them would occupy less space than a grain of sand. The malleability of gold—the perfume of musk—the odour of flowers—and many other instances might be given of the excessive minuteness of the atoms of matter: yet from a variety of circumstances it may be inferred that matter is not infinitely divisible."—*Mrs. Somerville*, p. 125.

† To those who feel desirous of pursuing this subject I would recommend the Natural History of Animalcules, by Andrew Pritchard, Esq. 1 vol. 8vo. 1834.

sea-weed or the rocks : this is the germ of a new
colony of this compound animal. As it increases
in magnitude, the usual character of the flustra may
be detected; and if the *fleshy* film be removed, a
spot of calcareous matter is left attached. In the
larger and free masses of flustra, the decomposition
of the animal substance after death is very manifest.
This specimen of *flustra foliacea*, which was dredged
up twenty miles SS.W. of Brighton, in eighteen
fathoms of water, and for which I am indebted to
my friend Robert Hannay, Esq., is a fine example
of this brittle species ; it was highly offensive
when first in my possession, from the emanations
evolved during the decomposition of the animal.
It is now a calcareous skeleton, with here and there
portions of the shrivelled integument, and of course
without any traces of polypi in the cells.

Let us now refer to the previous remarks, and
inquire if the flustra presents the essential characters
of animal existence. It possesses a determinate
form, and has a soft, fleshy substance, that can for
a certain period resist chemical and mechanical
agency. It is furnished with instruments capable of
moving with great celerity, susceptible of external
impressions, and expanding and contracting at will.
Here then is evidence of sensation and of voluntary
motion ; and although from the extreme minuteness
of the structure nerves cannot be detected, yet
there can be no doubt that the animal possesses a
nervous, and also a circulatory system for effecting

nutrition and reparation. We see also that when the flustra is removed from the element in which it lived, the substance of which it is composed, like the flesh of the larger animals, undergoes putrefaction—in other words, that the animal dies—it has lost the vital principle by which it previously resisted chemical agency, and now submits to the effects of those laws which act upon inorganic matter. The calcareous substance that formed its support or skeleton, and which, like the bones of mammiferous animals, was secreted by the fleshy mass, alone remains.

12. SKELETONS OF POLYPI, or CORALS.—I would here particularly remark, that the stony matter or support of all zoophytes is formed in the same manner; the specimens called corals being secretions from an animal substance by which they were invested, in like manner as the bones in man are secreted by the tissue or membrane designed for that purpose, and acting without his knowledge or control.   Nothing can be more erroneous than the common notion that the cells in the larger corals are built up by the polypi which are found in them, in the same manner as are the cells of wax by the bee or the wasp.

From what has been advanced, we perceive that the flustra is a compound animal, consisting of a fleshy substance, secreting and surrounding a calcareous skeleton, and studded over with cells or digestive sacs with polypi, which may be con-

sidered as foci of vitality, by whose agency the life
of the whole mass is maintained. Whether these
are separate centres of life, susceptible of pain and
pleasure independently of the whole, is impossible
to determine ; we have a living proof in our own
species in the Siamese twins, that there may be an
united organization with distinct nervous systems,
and individual sensations. However this may be,
we are at least certain that the Eternal has be-
stowed on these, as on all his creatures, the capacity
and means of enjoyment. In truth, when observing
the active movements of polypi, through the microscope of Mr. Lister, the beautiful remark of
Paley,* on the happiness enjoyed by animals, came
forcibly to my recollection, and I thought with
him, that if any motion of mute beings could ex-
press delight it was exhibited by the polypi on
which I was gazing : if they had intended to make
signs of their happiness they could not have done
it more effectually, for they were sporting about in
every direction, sometimes expanding like a flower,
then suddenly closing and partially retreating, and
again extending themselves to their extreme dimensions.

13. DIVERSITY OF FORM IN THE SKELETONS.—
In the flustra, then, we have the elements of zoophytal
organization, and the varied and extraordinary forms
which will hereafter come under our notice are but

* Natural Theology.

modifications of this type of animal existence.   In some, the skeleton or support consists of earthy matter, as in the flustra, but solid and hard as adamant; in many examples it branches out like vegetables; in others, constitutes hemispherical masses, covered by numerous convolutions somewhat resembling the brains of quadrupeds; and in some, forms an aggregation of tubes, terminating in star-like openings. Among the branched varieties, some are covered by pores so numerous as to be called millepora; in many, the openings are distant: some have stellular markings here and there; while in others, the whole surface presents a stellated structure.   In many species the fleshy animal matter entirely covers and conceals the stony skeleton during life; in others, the latter becomes exposed and forms a trunk, having branches covered by living polypi; while in another, and numerous division (of which the common *sertularia* is an example), the skeleton is secreted by the *outer* surface of the animal substance, and constitutes an external protection to the polypi.   In another family the skeleton is of a horny or ligneous texture, and flexible, as in the *Gorgonia*, bending to the motions of the waves; while in some it is jointed or articulated, as in the *Isis*.   Sometimes the skeleton is impressed with the cells, as in the madrepores; while in other species, as the red coral, the stem is smooth, and exhibits no traces of the peculiar structure of the animal.   Yet amidst these almost endless varieties

of form, the same essential characters are maintained ; in all there is a skeleton or solid support,
and a fleshy or gelatinous substance studded with
polypi.

From an analysis of the stony corals, it appears
that their composition is very analogous to that of
shells.   The porcellaneous shells, as the cowry, are
composed of animal gluten and carbonate of lime,
and resemble, in their mode of formation, the
enamel of the teeth ; whereas the pearly shells, as
the oyster, are formed of carbonate of lime and a
gelatinous or cartilaginous substance, the earthy
matter being secreted and deposited in the interstices of a cellular tissue, as in bones.   In like
manner some corals yield gelatine upon the removal of the lime ; while others afford a substance in every respect resembling the membranous
structure, obtained by an analysis of the nacreous
shells.*

14. GEOGRAPHICAL DISTRIBUTION OF POLY-
PARIA.—I will now consider the geographical distribution of these singular beings; in the next
place, describe a few of the principal varieties ;
and lastly, review the important physical changes
effected by creatures so minute, and apparently so
incompetent to produce any material alteration in
the earth's surface.

The greater number of the zoophytes or poly-

* Experiments of Mr. Hatchett.

paria are inhabitants of the ocean; many species prefer the immediate influence of atmospheric changes: they are seen on the rocks and plants which the tide leaves bare, sometimes in such profusion that the whole surface appears one animated mass; but most species suffer from the action of the air. At the period of the great equinoctial tides, when the sea retires from the rocks it has overflowed for many preceding months, the polypi, when the waters first recede, are full of vigour, but languish as they lose their moisture, nor fail to perish if they remain long uncovered by the sea.

Some species are situated on the southern slope of the rocks; others, on the contrary, are attached to the opposite aspect, and never to the former. The larger polyparia are rarely found in places exposed to violent currents; it is in the hollows of rocks, in submarine grottoes, in the shelter of large and solid masses, that these species attach themselves. Some appear fitted to enjoy the powerful action of the surges, their pliant branches bending to the movements of the waters, and floating in the agitated medium. Others form immoveable rocks, which increase slowly but surely, till they become elevated above the surface of the waters and constitute islands, as I shall hereafter describe.

The peculiarities in the distribution of these animals are not confined to the relative depths of the waters; like plants, they vary with the climate, and in cold latitudes the cellariæ and sertulariæ, with a few sponges and alcyonia, are alone to be met with.   As we proceed to the 44th or 45th degree of northern latitude, their number increases, and gorgoniæ, sponges with loose tissue, and mille-pores with foliated and fragile expansions, appear in profusion.   A little farther, and the coral reddens the depths of the ocean with its brilliant branches, and is soon followed by the large madrepores.*   It is not, however, before the 34th degree of northern latitude that the corals become developed to the grandeur and importance which they afterwards attain, to the extent of a parallel southern latitude. It is therefore within the tropics, in a zone of more than 60 degrees expansion, that these beings, scarcely visible to the naked eye, exercise their empire in a medium whose temperature knows no change.   From the depths of the ocean they elevate those immense reefs that may hereafter form a communication between the inhabitants of the tem-perate zones.†   I proceed to notice a few of the principal forms of polyparia.

15. THE FLUSTRA, or SEA-MAT. (Pl. 1, figs. 2, 7, 12; Pl. 2, fig. 9.)—My previous observations on the structure of this genus of zoophytes render

---

* Lamouroux.

† CORALLINA; an excellent Abstract of Lamouroux's Me-moir on the Flexible Corals. One vol. 8vo. 1834. This work is so scarce, that a new edition is in great request.

further details unnecessary. The Flustræ present great variety of form, sometimes being attached to marine plants, which they inclose, as it were, in a living sepulchre; at others, spreading into thin foliated expansions, which have both sides studded with cells. The prevailing colour is white, or a light fawn, but some species have a tinge of pink or yellow. They abound in every sea, and are not restricted by climate; occur in profusion along the sea-shores, and are found attached to the fuci that are thrown up from the profound depths of the ocean. The small parasitical species, when dried, appear like spots of a chalky substance on the sea-weed (see Pl. 1, fig. 12). The increase of the Flustra is thus described by Lamouroux:*—"When the animal has acquired its full growth, it flings from the opening of its cell a small globular body, which fixes near the aperture, increases in size, and soon assumes the form of a new cell; it is yet closed, but, through the transparent membrane that covers its surface, the motions of the polypus may be detected; the habitation at length bursts, and the tentacula protrude, eddies are produced in the water, and conduct to the polypus the atoms on which it subsists."† The Flustræ are very abundant in a fossil state; there is scarcely an echinus

* Corallina, page 43.
† Dr. Grant's interesting Observations on the gemmules of the Flustra are given by Dr. Roget, in the Bridgewater Essay, page 169.

of the chalk that has not several parasitical species attached to its shell.

16. SERTULARIÆ, or VESICULAR CORALLINES. (Pl. 2, figs. 6, 8.)—The elegant arborescent forms of the Sertulariæ, must be familiar to every one who has rambled by the sea-side. This branch of the sea-pine coralline, (*sertularia pinaster*, Pl. 2, fig. 8,) which is shown magnified in this sketch, (fig. 8 *a*,) exhibits the usual appearance of these corallines. The sertulariæ consist of tubes united together, and having lateral apertures for the protrusion of each polype; one elegant species, the *sertularia setacea*, is very abundant on the shores at Brighton after storms, being attached to the seaweed. This representation of a branch magnified sixty times, shows the form of the polypi,* which, when fully expanded, are of great beauty. On one occasion, when I was present, Mr. Lister was observing a specimen of this creature, when a little globular animalcule swam rapidly by one of the expanded polypi; the latter instantly contracted, seized the globule, and brought it to the mouth or central opening by its tentacula; these gradually opened again, with the exception of one which remained folded, with its extremity on the animalcule. The mouth instantly seemed filled with hairs, that closed over the prey, which, after a few seconds, was carried slowly down, the

---

* In Pl. 2, fig. 6, the number of tentacula is too small; each polype has from 16 to 20.

mouth contracting and the neck enlarging into the stomach; here it was imperfectly seen, and soon disappeared.*

The *campanulariæ*, so named from their bell-shaped cells placed on foot-stalks, are also abundant on these shores. Pl. 1, fig. 4, is a magnified view of a branch of campanularia with two cells; in one the polype is expanded, in the other contracted. Viewed alive through the microscope, currents of minute globules are seen constantly running along the tubes, and are probably induced by the action of invisible cilia.

17. GORGONIA; SEA-FAN or FEATHER.—The Gorgonia Flabellum, or Venus's Fan, is a common flexible coralline, an inhabitant of almost every sea, and frequently attaining a height of four or five feet. When fresh from the water it is of a bright yellow colour. This species exhibits the usual structure of the corticiferous polyparia, or zoophytes which are composed of two substances, namely, an internal axis or skeleton of a tough horny consistence, and an external envelope or rind, which entirely invests the former. The Gorgoniæ present great diversity of form and appearance. This specimen from the West Indies, (Pl. 2, fig. 2,) is remarkable for its richness of colour, being a bright yellow, spotted with red; this species, (Pl. 2, fig. 4,) from the Mediterranean, has its pendant branches very

* Philos. Trans. 1834, p. 372.

elegantly disposed, and is of a purplish-lake colour ; in both these examples the axis is black, and of the consistence of tough horn.   Another beautiful species from the Mediterranean, the *gorgonia patula* of Ellis, is of a bright red, and has the openings for the polypi disposed in two rows ; a portion, highly magnified, is here represented, (Pl. 1, fig. 2,) and exhibits two polypi protruding ; one closed, the other expanded.

These flexible polyparia are attached to the rocks by an extended base, whose surface is usually deprived of the fleshy substance by which the other parts are invested.   The stem which springs from the base, although in a few species simple, generally divides into branches, which are exceedingly various in their size and distributions ; double, single, anastomosed, pinnated, straight, and pensile ; and the stems are either compressed, flat, angular, or cylindrical ; but in all these modifications the same general structure prevails—an axis, and an external crust or rind.   The former is either horny, elastic, flexible, brittle, or pithy, and of a dark colour ; the latter a soft fleshy substance, studded with pores, from whence the polypi issue when the animal is alive ; this rind becomes earthy and friable when dried.   In the water the various species present the most vivid hues of red, green, violet, and yellow.   The Gorgoniæ are found in every sea, but certain species appear to be restricted to hot regions ; all inhabit deep water.   I believe but few traces of this genus have been found in a fossil state.

18. The Red Coral; *corallium rubrum.* (Pl. 1, figs. 1, 6.)—I advance to the examination of the polyparia whose axis is composed of a calcareous stony substance; and one genus of which possesses a skeleton of such beautiful colour, and susceptible of so high a polish, as to be eminently employed for purposes of ornament. The red coral is a branched zoophyte, somewhat resembling in miniature a tree deprived of its leaves and twigs. It seldom exceeds one foot in height, and is attached to the rocks by a broad expansion or base. It consists of a bright red, stony axis, invested with a fleshy, or gelatinous substance of a pale blue colour, which is studded over with stellular polypi. This figure (Pl. 1, fig. 1,) represents a branch of the natural size; *a a*, the extremities, covered with the corticiferous substance, and having many polypi expanded; *b*, the axis, or internal skeleton, deprived of its rind; fig. 6, a fragment of a branch, with several polypi, magnified. As the polypiferous centres are composed of the animal crust which undergoes decomposition, no traces remain of their structure on the durable skeleton.

The interior of the red coral, as is well known, is so dense and compact as to be susceptible of a high polish, and forms an important article of commerce: it is obtained by dredging in different parts of the Mediterranean and Eastern seas. It varies much in its hue, according to its situation in the sea: in shallow water it is of the most beautiful colour, a free admission of light appearing necessary for the full development of its energies. It is of slow growth, eight or ten years, in a moderate depth of water, being necessary for it to reach maturity. Arrived at this period it extends, though very slowly, and is soon pierced on all sides by those destructive animals which even attack the hardest rocks; it loses its solidity, and the slightest shock detaches it from its base. Becoming the sport of the waves, the polypi perish, their brilliant skeleton is exposed, and thrown upon the shore; its bright hue soon disappears, and it is reduced to fragments by the attrition of the waves, or, mixed with the remains of shells and other marine exuviæ, and thrown up by the tides, or drifted inland by the winds, it assists in forming those accumulations of the spoils of the sea which constitute the modern conglomerates described in a previous lecture, (pp. 61, 67.)

19. TUBIPORA; *Organ-pipe coral.* (Pl. 2, figs. 10, 10 *a.*)—This genus of corals is well known, from the elegance and beauty of one species, (*Sarcinula musicalis,*) which is common in most collections. In this fine specimen, from the Rev. T. Trocke, you perceive that the coral is composed of parallel tubes, united by lateral plates, or transverse partitions, placed at regular distances, (Pl. 2, fig. 10;) in this manner large masses, consisting of a

congeries of pipes or tubes, are formed. When the
animal is alive, each tube contains a polypus of
a beautiful bright green colour, and the upper
surface of the coral is covered with a gelatinous
mass from the confluence of the polypi; a magnified view, with a polypus and section of two other
tubes, is here represented, (Pl. 2, fig. 10 *a.*) This
species occurs in great abundance on the coast of
New South Wales, in the Red Sea, and in the
Molucca Islands, varying in colour from a bright
red to a deep orange. It grows in the shape of
large hemispherical masses, from one to two feet in
circumference; these first appear as small specks
adhering to a shell or rock; as they increase, the
tubes resemble a group of diverging rays, and at
length other tubes are produced on the transverse
plates, thus filling up the intervals, and constituting
a uniform tubular mass; the surface being covered
with a green fleshy substance, studded with stellular polypi.

20. MADREPORES ; *madrepora corimbosa.* (Pl. 1,
fig. 5.)—In the red coral, no cells are formed on
the skeleton to serve as a protection to the polypi;
but in another family of branched, or arborescent,
calcareous polyparia, little cups or cells, with radiating lamellæ, are composed of the substance of the
skeleton, and in which the polypi are situated.
When the animal dies, and the outer carneous investment is decomposed, the axis is seen to be studded
over with elegant, lamellated cells, or stars, variously

formed and arranged, in different species. The
white branched corals, commonly seen in collections,
belong for the most part to madrepores; it is not,
therefore, requisite to describe this form of zoophyte more in detail.   In the water the madrepores
are invested with a fleshy integument of various
colours; and each cell has a polypus similar to the
corals I have described.   This representation (Pl. 1,
fig. 5,) of a madrepore, as seen alive in the water,
will serve to convey a general idea of the nature of
the original; from each of the projecting cells the
tentacula of a polypus issued, and the branches
seemed alive with their hydra-like forms.

21. THE ACTINIA, or SEA ANEMONE.   (Pl. 2,
figs. 12, 12 a.) — In another division of corals the
cells are few and of considerable dimensions, the
polypi being of proportionate size, and bearing
considerable analogy to the actiniæ, or sea-anemones; a few observations on these animals will
therefore enable us to comprehend the nature of the
polyparia next to be considered.   The Actinia, or
sea-animal flower, as it is termed, appears, when
quiescent, like a mass of tough jelly, of a subcylindrical form, and of various tints of crimson,
green, blue, brown, or mottled; see Pl. 2, fig. 12;
when expanded it presents a broad disk, surrounded
by tentacula, having in the centre a corrugated
surface, which is contracted into a marsupial or
purse-like form.   The actiniæ are attached to the
rocks by a broad base, but they can detach

themselves, and change their position; on this coast
hundreds may be seen, at low water, in the hollows
of the chalk, left bare by the reflux of the tide.
They are carnivorous and very voracious, feeding
on the small fish, crustacea, or mollusca, that come
within their reach. I have kept them for months
in sea-water, supplying them daily with meat, which
they greedily seized, drew into their sac, and afterwards ejected perfectly colourless, having absorbed
the juices, and left the tough, muscular fibre. The
body of the actinia is highly contractile, and full
of cells; the tentacula (see Pl. 2, fig. 12 a,) are
hollow tubes, which the animal has the power of
filling with sea-water, and thus causing them to
protrude; a mechanism similar to that of the spiral
appendages of the terebratulæ described in the
former lecture (page 422.). The cells also contain
water, with which the whole or any part of the
body can be filled; and I have observed, that when
the animal was desirous of shifting its situation, it
distended one half of the body with water, then
withdrew the base from the stone, and sank to the
bottom of the vessel in which it was contained.
The surface of the stomach, and even the internal
lining of the tentacula, are abundantly furnished
with cilia. The actinia has no durable skeleton.

22. CARYOPHILLIA. TURBINOLIA, &c. (Pl. 1,
figs. 9, 13; Pl. 2, figs. 3, 11.)—In this small coral
(*caryophillia cyathus*), from the Mediterranean,
(Pl. 2, fig. 11,) we have an isolated calcareous cell,

divided by vertical lamellæ placed in a radiating manner. This cell is the skeleton of a single polypus, having a double row of tubular tentacula, and bearing a great analogy to the actinia; indeed, the recent animal may be described as an actinia with a calcareous skeleton, fixed by its base. This specimen, (Pl. 2, fig. 3,) is another species of a similar polypus. In the caryophilliæ possessing more than one cell, each receptacle contains a polypus, as in the *caryophillia angulosa*, (Pl. 1, fig. 13,) which is here represented as seen alive in the water. In another genus, *pocillopora*, (Pl. 1, fig. 9,) the investing fleshy skin is beautifully mottled, and the polypi are terminal as in the caryophilliæ.

23. FUNGIA. (Pl. 1, fig. 3; Pl. 2, fig. 5.)—The white, disciform, lamellated corals, called sea-mushrooms, from their fancied resemblance to fungi, are among the most elegant forms of polyparia in the cabinets of collectors. These, in a living state, are covered with a thick fleshy substance, transparent like jelly, which fills up all the numerous radiating interstices of the calcareous laminæ; see Pl. 1, fig. 3; in the central depression the fleshy mass is formed into a large polypus with tentacula; in the fungia there is but one polypus—but one focus of vitality. In the fungia actiniformis, (Pl. 2, fig. 5,) the polypus strikingly resembles the actinia; and the whole surface of the disk is covered with long, tubular, conical, prehensile tentacula, with minute terminal apertures, and striated, transverse, muscular bands;

these tentacula are protruded by the injection of water from below, as in the actinia. In the fungia the stony base is secreted from the inferior surface of the soft substance, and attached or cemented, as it were, to the rock.

24. ASTREA, PAVONIA, &c. (Pl. 2, figs. 1, 7, 7^a.) —In some of the large massive lithophytes, the cells for the polypi are very numerous, and the coralline mass presents a surface beautifully marked with stellular impressions. The astrea viridis is here represented as seen alive in the sea, (Pl. 2, fig. 7;) in one part the fleshy investment is removed to show the calcareous cells (g, n). The polypi in this species are of a dark-green colour, more than six lines in length (a, b, c), and are protected by deep, laminated, polygonal cells, two lines in diameter. They are striated with longitudinal and transverse bands, (fig. 7, b, fig. 7^a, d), and connected by a fleshy layer (fig. 7, f,) which covers the dark-brown coral; some of the polypi are seen expanded, and others contracted. In the magnified view, fig. 7^a, of a single polypus, the tentacula are shown in an expanded state, disposed around the prominent blue mouth. The appearance of groups of astreæ, and other corals, is described as being most beautiful when viewed with the animals alive and in activity; looking down through the clear sea-water, the surface of the rocks appears one living mass, and the polypi present the most vivid hues.

The Pavoniæ are those corals which have deep

and isolated cells, each containing a large depressed polypus, very similar in its appearance and structure to the actinia.   Pl. 2, fig. 1, is a representation of a single cell and polypus of the *P. lactuca*, from the shores of the South Sea islands.   The polypi are of a deep-green colour, and there is a connecting, transparent, fleshy substance, which extends over the extreme edges of the foliated expansion of this elegant coral.   " From the magnitude and muscularity of the polypi in the large lithophytes, and the increased number and strength of their prehensile organs, they are capable of seizing and digesting more highly organized prey than the delicate, minute, cellular forms of the flustræ." *

25. MEANDRINA *cerebriformis ;* or brain-coral.— The large hemispherical corals, whose surface is covered over with meandering ridges and depressions, disposed in a manner that somewhat resembles the convolutions of the brain, are well known by the name of *brain-stone.*   In a living state the mass is covered with a fleshy substance, variously coloured, and having numerous short, conical, polypiform orifices.   This sketch, (Pl. 1, fig. 11,) shows the appearance of the coral in the water; the polypi are retracted and concealed.   This coral sometimes attains considerable magnitude; a very beautiful specimen in the British Museum is four feet in circumference.   The base of the meandrina,

* Dr. Grant's Comparative Anatomy.

like that of the fungia, is adherent to the rock, with which, being formed of a like material, it becomes identical. As one fleshy mass expires, another appears, and gradually expands, pouring out its calcareous secretion on the parent mass of coral; thus successive generations go on accumulating vast beds of stony matter, and lay the foundation for coral reefs and islands. We may compare, observes Mr. Lyell,* the operation of the zoophytes in the ocean, to the effects produced on a smaller scale on land, by the plants which generate peat; in which the upper part of the *sphagnum* (page 41) vegetates, while the lower is entering into a mineral mass, in which the traces of organization remain when life has entirely ceased. In corals, in like manner, the more durable materials of the generation that has passed away, serve as the foundation on which their progeny are continuing to spread successive accumulations of calcareous matter.

26. APPEARANCE OF THE LIVING CORALS IN THE SEA.—In some parts of the sea the eye perceives nothing but a bright sandy plain at bottom, extending for many hundred miles; but in the Red Sea, the whole bed of this extensive basin of water is absolutely a forest of submarine plants and corals. Here are sponges, madrepores, corals, fungiæ, and other polyparia, with fuci, algæ, and all the variety of marine vegetation, covering every part of the

* Principles of Geology.

bottom, and presenting the appearance of a submarine garden of the most exquisite verdure, and
enamelled with animal forms, resembling, and even
surpassing in splendid and gorgeous colouring, the
most celebrated parterres of the East.

Ehrenberg, the distinguished German naturalist,
whose labours have so greatly advanced our knowledge of the infusoria, was so struck with the magnificent spectacle presented by the living polyparia
in the Red Sea, that he exclaimed with enthusiasm,
" Where is the paradise of flowers that can rival in
variety and beauty these living wonders of the
ocean?"   Some have compared the appearance to
beds of tulips or dahlias ; and, in truth, the large
fungiæ (Pl. 1, fig. 3, and Pl. 2, fig. 5), with their
crimson disks, and purple and yellow tentacula, bear
no slight resemblance to the latter.

Instead of one lecture, many would be required fully
to elucidate the natural history of the polyparia, so
numerous and so varied in appearance and structure
are the modifications of this class of animal existence.

27. CORAL REEFS.—I have already alluded to
the vast accumulations of calcareous rocks in tropical
seas, resulting from the consolidation of the skeletons
of polyparia ; but the physical changes that are
produced by such apparently inadequate means
require farther consideration, since they illustrate
the formation of the coralline rocks, which will hereafter come under our notice.

In the flustra foliacea of our coast (page 460),

delicate and brittle though it be, we may perceive the elements of those important changes to
which the large lamellar polyparia of tropical seas are
giving birth. In the specimen before us, you may
observe that the base of the mass of flustra, which is
about six inches in diameter, is already consolidated
by an aggregation of sand, which has filled up the
interstices. On the surface are numerous parasitical
shells and corals; between the convolutions of its
foliated expansions, echini, crustacea, and other
animals, have taken shelter; sand and mud have
invested every cranny of the lower third of the
specimen, and imbedded serpulæ, sabellæ, and fragments of many species of shells. It is evident,
that were the whole specimen filled up and surrounded by such detritus, as it shortly would be in
its native element, a solid block would be formed,
exhibiting, when broken, the remains of the flustra,
impacted in a conglomerate of sand, shells, and
corals. Thus we perceive that even the delicate,
friable skeleton of the flustra of our shores, may form
the nucleus of a solid rock; and in the process I
have described, we have, as it were in miniature, the
formation of a coral reef.

28. CORAL REEF OF LOO CHOO.—But it is in
tropical seas that the meandrinæ, astreæ, caryophilliæ, and other stony corals, form those immense
masses, which not only give rise to groups of islands
in the bosom of the ocean, but are gradually forming tracts of such extent, that a new continent may

spring up where the fabled Atalantis once flourished.
From the many interesting descriptions of the nature
and formation of coral reefs and islands that have
been published by our voyagers, I select the following graphic account, by Captain Basil Hall, of a
coral reef near the great island of Loo Choo:—

"When the tide has left the rock for some time
dry, it appears to be a compact mass, exceedingly
hard and rugged: but as the tide rises, and the
waves begin to wash over it, the polypi protrude
themselves from holes which were before invisible.
These animals are of a great variety of shapes and
sizes, and in such prodigious numbers, that in a
short time the whole surface of the rock appears to
be alive and in motion. The most common form is
that of a star, with arms, or tentacula, which are
moved about with a rapid motion in all directions,
probably to catch food. Others are so sluggish
that they may be mistaken for pieces of the rock,
and are generally of a dark colour. When the
coral is broken about high-water mark, it is a solid
hard stone; but if any part of it be detached at a
spot where the tide reaches every day, it is found to
be full of polypi of different lengths and colours;
some being as fine as a thread, of a bright yellow,
and sometimes of a blue colour. The growth of
coral appears to cease when the worm is no longer
exposed to the washing of the sea. Thus a reef
rises in the form of a cauliflower, till the top has
gained the level of the highest tides, above which

the animalcules have no power to advance, and the
reef of course no longer extends upwards."

29. CORAL ISLANDS.—Kotzebue, Flinders, and
MM. Quoi and Gaimard, have severally described
the formation of coral islands ; the following is an
abstract of their observations :—

The coral banks are everywhere seen in different
stages of progress : some are become islands, but
not yet habitable; others are above high-water
mark, but destitute of vegetation ; while many are
overflowed with every returning tide. When the
polypi, which form the corals at the bottom of the
ocean, cease to live, their skeletons still adhere to
each other, and the interstices being gradually filled
up with sand and broken pieces of corals and shells,
washed in by the sea, a mass of rock is at length
formed. Future races of these animalcules spread
out upon the rising bank, and in their turn die,
increase and elevate this wonderful monument of
their existence.

The reefs which raise themselves above the level
of the sea, are usually of a circular or oval form,
and surrounded by a deep and oftentimes unfathom-
able ocean. In the centre of each there is generally
a shallow lagoon, with still water, where the smaller
and more delicate kinds of zoophytes find a tranquil
abode ; while the stronger species live on the outer
margin of the isle, where the surf dashes over them.
When the reef is dry at low water, the coral animals
cease to increase. A continuous mass of solid stone

is then seen, which is composed of shells and echini, with fragments of corals, united by calcareous sand, produced by the pulverization of the shells of friable polyparia.   Fragments of coral limestone are thrown up by the waves; these are cracked by the heat of the sun, washed to pieces by the surge, and drifted on the reef.   After this the calcareous mass is undisturbed, and offers to the seeds of the cocoa, pandanus, and other trees and plants, floated thither by the waves, a soil on which they rapidly grow, and overshadow the white, dazzling surface.   Trunks of trees, brought by currents from other countries, find here at length a resting place.   With these come some small animals, as lizards and insects.   Even before the trees form groves or forests, sea-birds nestle there; strayed land-birds find refuge in the bushes; and at a still later period, man takes possession of the newly created country.   It is in this manner that the Polynesian Archipelago has been formed.   The immediate foundations of the islands are ancient coral reefs, and these, in all probability, are based on the cones or craters of submarine volcanoes long since extinct.   There is another circumstance worthy of remark : most of these islands have an inlet through the reef opposite to the large valleys of the neighbouring land, whence numerous streams issue and flow into the sea ; an easy ingress is thus afforded to vessels, as well as the means of obtaining a supply of fresh-water.

Of the grand scale on which the changes here contemplated are going on, we may form some idea from the facts stated by competent observers, that in the Indian Ocean, to the south-west of Malabar, there is a chain of reefs and islets 480 geographical miles in length; on the east coast of New Holland, an unbroken reef 350 miles long; between that and New Guinea, a coral formation which extends upwards of 700 miles; and that Disappointment Islands and Duff's Group are connected by 600 miles of coral reefs, over which the natives can travel from one island to another.

30. MONTGOMERY'S DESCRIPTION OF CORAL ISLANDS.—There is so much of the marvellous and sublime in the idea of the creation of islands and continents by the ceaseless labours of countless myriads of unconscious living instruments, that we cannot be surprised that this interesting subject has attracted the attention of one of the most amiable and elegant of our modern poets; I will relieve this detail by the following beautiful extract from the Pelican Island, of James Montgomery :—

> " I saw the living pile ascend,
> The mausoleum of its architects,
> Still dying upwards as their labours closed;
> Slime the material, but the slime was turned
> To adamant by their petrific touch.
> Frail were their frames, ephemeral their lives,
> Their masonry imperishable. All
> Life's needful functions, food, exertion, rest,

By nice economy of Providence,
Were overruled, to carry on the process
Which out of water brought forth solid rock.
Atom by atom, thus the mountain grew
A coral island, stretching east and west;
Steep were the flanks, with precipices sharp,
Descending to their base in ocean gloom.
Chasms few, and narrow, and irregular,
Formed harbours, safe at once and perilous—
Safe for defence, but perilous to enter.
A sea-lake shone amidst the fossil isle,
Reflecting in a ring its cliffs and caverns,
With heaven itself seen like a lake below.
Compared with this amazing edifice,
Raised by the weakest creatures in existence,
What are the works of intellectual man,
His temples, palaces, and sepulchres?
Dust in the balance, atoms in the gale,
Compared with these achivements in the deep,
Were all the monuments of olden time—
Egypt's grey piles of hieroglyphic grandeur,
That have survived the language which they speak,
Preserving its dead emblems to the eye,
Yet hiding from the mind what these reveal;
Her pyramids would be mere pinnacles,
Her giant statues, wrought from rocks of granite,
But puny ornaments for such a pile
As this stupendous mound of catacombs,
Filled with dry mummies of the builder worms."

31. FOSSIL ZOOPHYTES.—Although many genera
of polyparia are omitted in this brief sketch, I
must pass without farther remark to the consideration of the fossil zoophytes, of which, in this
place, I purpose offering a cursory review.   The

formation of conglomerates, from the debris of corals and shells, has been fully explained in a former lecture, (page 60.) In the newer pliocene of Palermo many Mediterranean corals are imbedded. The blocks of silex, agate, and chalcedony, which are scattered over some districts in the West Indies, frequently contain meandrinæ, astreæ, and caryophilliæ, in a silicified state ; and polished slices exhibit the internal structure of the corals in a highly beautiful manner, as may be seen in these specimens, presented me by the Hon. Mrs. Thomas. The Crag (page 190) abounds in flustræ, sertulariæ, and other genera of the small polyparia, apparently of species that inhabit the British seas. In the older tertiary the remains of turbinoliæ, caryophilliæ, astreæ, fungiæ, and of other genera, amounting to about thirty species, have been discovered.

32. ZOOPHYTES OF THE CHALK.—In the chalk formation the remains of this family occur in profusion. In the upper divisions, the Maestricht beds, the lamelliferous corals, as the astreæ, meandrinæ, fungiæ, &c. prevail, and may be extracted from the friable arenaceous strata, in a fine state of preservation; the celluliferous genera are equally abundant. In the white chalk the large calcareous polyparia are rare, while the fibrous zoophytes occur in profusion, both in the chalk and in the nodules of flint. Sponges of several kinds, and many species of allied genera, abound in every chalk-quarry in the south-east of England. One

of the most common of the poriferous zoophytes
is a species which occurs either in an expanded
state like a broad flat disk, or contracted into
a cyathiform shape ; the latter, when silicified,
gives rise to flints, which from their forms have
acquired the name of petrified mushrooms, or
goblets, according as the cavity is either full or
empty.   I have, in the " Fossils of the South
Downs,"* so fully explained the structure of the
original, under the name of *Ventriculite,* that it will
here be sufficient to observe that the living zoophyte
must have been of a cyathiform figure, and composed of a tough, jelly-like substance, capable of
expansion and contraction.   The smaller extremity
was attached to the rock by root-like processes ;
the outer tissue consisted of a net work of cylindrical,
perhaps tubular, fibres ; the inner surface of the
funnel-like cavity was studded with polypiferous
cells or openings.   Silicious specimens occur in
every variety of shape which such a structure could
assume ; some are conical and hollow, others resemble fungi, many are turbinated, and not a few
appear like a flat disk or plate ; in all, the margins
are more or less impressed with undulating lines,
which are produced by the section of the inclosed
zoophyte.

The *Choanite,* called petrified sea-anemone by
lapidaries, (page 288,) bears a close analogy to

* Illustrations of the Geology of Sussex, p. 167.

the recent alcyonia, which are zoophytes of a fleshy or gelatinous substance, invested with a tough outer skin, the surface of which is covered with pores. The alcyonia are permanently fixed by the base, and imbibe and eject sea-water, after the manner of the sponges. In these creatures we have the rudiments of a skeleton, for many species have acicular, silicious spines; hence the name of sea-nettles given to those varieties which wound or sting on being handled. The spines are of various forms, as you may observe in this enlarged representation of three varieties, (Pl. 1, fig. 10 a, 10 b, 10 c.) In the choanite, crucial spines, resembling those in the recent alcyonia, may be detected. The Rev. J. B. Reade has ascertained that in some flints the external layer is almost wholly composed of the spines of poriferæ. The choanite is of a subcylindrical form, with root-like processes, and having a cavity or sac which is deep, and small in comparison to the bulk of the animal. The inner surface is studded with pores, which are the terminal openings of tubes, disposed in a radiating manner, and ramifying through the mass. The beautiful markings observable in many pebbles, collected on the shores at Brighton and Bognor, are derived from the internal structure of this zoophyte;* the horizontal sections display a central disk, from

* See Geology of the South-East of England, p. 106. Thoughts on a Pebble; the frontispiece of this little work epresents a pebble inclosing a choanite.

whence the tubes diverge to the circumference—
the vertical, an elongated conical space, with
lateral rays. Ramose sponges form the nuclei of
most of the irregular branched flints; and zoophytes related to other genera of poriferæ are
equally abundant. A small species of caryophillia
occurs in the chalk of Sussex, and a turbinolia in
the Galt.

33. ZOOPHYTES OF THE SHANKLIN SAND. In
the arenaceous strata of the chalk formation, the

TAB. 56.—FOSSIL ZOOPHYTE FROM FARINGDON.

Chenendopora fungiformis.

Shanklin Sand, immense numbers of zoophytes, par-
ticularly of the poriferæ, occur in some localities.
The gravel-pits in the immediate neighbourhood of
Faringdon, in Berkshire, are extremely prolific in

these remains. The beds consist of an aggregation of sand, and pulverized shells and corals, impregnated with iron, and containing myriads of shells, polyparia, and casts of nautili and ammonites. The great mass of the materials is in the state of a loose conglomerate, but here and there indurated blocks occur. In a few visits to these quarries I collected numerous specimens of nautili, belemnites, ammonites, ostreæ, terebratulæ, echinites, and their spines; milleporæ, tubiporæ, and several minute corallines ; spongiæ, alcyonia, lymnoreæ, and many species that are undescribed. One of the most common and perfect of the porifera of the Faringdon pits, is an elegant cyathiform zoophyte (Tab. 56), called by the quarrymen *petrified salt-cellar.*

The sand near Warminster has yielded to the researches of Miss Etheldred Benett, of Norton House, (a lady, whose works have so ably illustrated the Geology of Wiltshire,) many species of fibrous zoophytes, of which but few traces occur in the sand of Kent or Sussex.

In the Shanklin sand of the Isle of Wight, Mr. Webster long since discovered the remains of a zoophyte, which from its form, he called the tulip alcyonium.* The original appears to have resembled in shape a closed tulip, having a pyriform bulb or head supported by a long stem, with a broad base for attachment to the rock. Sections of the bulb

* Syphonia pyriformis of Goldfuss. Dr. Fitton's Memoir, Plate XV.

display a congeries of longitudinal tubes, arranged somewhat concentrically; and these may be traced from the base and along the stem. The structure of the original animal appears to have borne an analogy to that of the alcyonia; the fossils are generally silicified throughout. A species of syphonia abounds in the silicious nodules of the chalk near Lewes and Brighton; and in one quarry, every flint contains vestiges of this kind of zoophyte.

The manner in which the remains of polyparia are distributed in the chalk involves an interesting inquiry; they occur promiscuously intermingled with shells, echini, and fishes; we find no beds of corals—nothing to point out the former existence of reefs. This phenomenon, however, is in accordance with the lithological characters of the chalk formation, and the nature of its organic remains; both of which indicate a profound ocean. As polyparia can only exist at moderate depths, the occurrence of reefs or beds of corals was not to be expected. About 150 species of zoophytes have been discovered in the cretaceous deposites; and entire strata of chalk are composed, like the modern calcareous beds of the Bermudas, of the detritus of polyparia.

34. CORALS OF THE OOLITE.—The Oolite, as I have previously remarked, abounds in corals, and contains beds which are decidedly coralline reefs, and have undergone no change but that of elevation from the bottom of the deep, and the consolidation

of their materials (page 389). The coral-rag of the
Oolite presents all the characters of a modern bank ;
the polyparia belong to astreæ, caryophilliæ, (Tab.
57, fig. 6,) madreporæ, meandrinæ, and other
genera which principally contribute to the formations
now going on in the Pacific. Shells, echini, teeth
and bones of fishes, and other marine exuviæ
occupy the interstices between the corals, and the
whole is consolidated by sand and gravel, held
together in some instances by calcareous, in others
by silicious infiltrations.* The corals, shells, &c.
are of species not now known in a living state.
Those who have visited the district where the coral-
rag forms the immediate sub-soil, and is exposed to
view in quarries, or in natural sections, must have
been struck with the almost identical features presented by these strata and the modern coral-banks.
We know that in our present seas all situations and
circumstances are not alike favourable to the existence and growth of polyparia ; in some parts of the
ocean they abound, and in others are altogether
wanting. In like manner in that enormous series of
deposites, the Oolite or Jura formation, which, as
we have seen, extends over a great part of Europe,
and has been formed in a vast sea, coral beds are
not universally distributed, but occur only in certain
localities ; in other terms, they are found to occupy

* At Tisbury, in Wiltshire, a beautiful silicified coral occurs,
a polished section of which is represented, Tab. 57, fig. 9 ; no
trace of the calcareous earth of the coral remains.

K K 2

the situations which in their native seas presented the conditions required by their peculiar organization. The zoophytes of the Oolite have yielded about 200 species, all of which are extinct.

In the lias but few polyparia are preserved; and the saliferous system of deposites presents but three or four species of gorgonia, three of retepora, and one of astrea.

35. CORALS OF THE OLDER SECONDARY.—The mountain limestone of the carboniferous system, which will be described in the next lecture, abounds in the cellular and lamelliferous zoophytes. In the early transition rocks, entire beds are composed of the remains of polyparia; and a few species of corals constitute the last trace of animal organization in the crust of our globe.

The simple turbinated corals having, like the fungia, but a solitary cell, inhabited by one polypus, occur in the limestones of Dudley in great abundance and perfection (Tab. 57, figs. 1, 3); a small species of fungia (Tab. 57, fig. 2) is also found associated with the immense mass of marine exuviæ of which those strata are composed.

The tubiporæ are among the earliest traces of organic bodies that are observable in the ancient strata. Many of the limestones are an aggregation of an extinct species of this genus; and on masses long exposed to the weather* (Tab. 58, fig. 1), the

* I am indebted for specimens to Sir George Sitwell, Bart. and J. Meynell, Esq.

structure of the original coral is sometimes well displayed, the tubes appearing in relief on the surface
of the rock.

TAB. 57.—FOSSIL CORALS.

Fig. 1. *Turbinolia Parkinsoni.*   2. *Fungia.*   3. *Turbinolia.*   4. *The
surface of Cyathophyllum hexagonum.*   5. *Syringopora geniculata of
Professor Phillips.*   6. *Caryophillia from the Oolite.*   7. *Milleporite.*
8. *Cyathophyllum basaltiforme.*   9. *Silicified astrea, from Tisbury.*
10. *Coralline marble.*

36. CORALLINE MARBLES.—A reddish marble,

beautifully marked by the sections of the inclosed tubipores, and susceptible of a good polish, is quarried in some parts of Derbyshire. (Tab. 58, fig. 2).

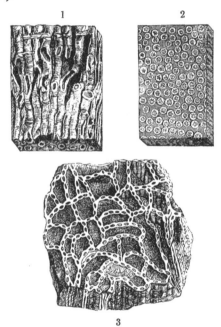

TAB. 58.—FOSSIL TUBIPORES AND CORALLINE MARBLE.

Fig. 1. *Tubipore in Derbyshire limestone.* 2. *Marble composed of tubipores.* 3. *Chain-coral (Catenipora labyrinthica) from Dudley.*

Mr. Parkinson has ascertained that the hue of this marble is dependent on the original colour of the coral, which, like the recent tubipore (page 478), was probably of a rich scarlet.

I have mentioned (page 467), that the earthy matter of the recent corals, like the phosphate of lime in the bones of animals, is formed by secretion from a membraneous structure, and that the lime may be removed by a chemical process, and the membrane rendered manifest. Few, however, will be prepared to learn, that even in corals, which have for countless ages been entombed in the solid rock, the animal tissue can be detected. To my late friend, Mr. Parkinson, we are indebted for the knowledge of this interesting fact. He immersed a piece of tubiporitic marble (Tab. 58, fig. 2) in diluted muriatic acid, which has the property of dissolving calcareous earth, but cannot affect animal matter : to employ his own words, " As the calcareous earth dissolved, and the carbonic acid gas escaped, I was delighted to observe the membraneous substance depending from the marble in light flocculi, of a deep red colour; and although not retaining the tubular form of the original coral, yet appearing in a beautiful and distinct manner."

A curious form of tubipore occurs in the Dudley limestone, which from the appearance of sections of the tubes, has received the name of chaincoral (Tab. 58, fig. 3). The tubes of this species are oval, and arranged perpendicularly side by side, in undulating lines; transverse sections, therefore, give rise to elegant markings on the surface, resembling delicate chain-work. This tubipore is only found in the ancient rocks.

The pebbles that are thrown by the waves upon the shore near Torquay, on the Devonshire coast, and which are water-worn fragments of the limestones of the country, disclose remains of corals; and when cut and polished, exhibit sections of the structure of the original polyparia.*

The ornamental marbles of Babbicombe, Bristol, &c., owe their beauty to the inclosed zoophytes. The black Kilkenny marble, in such general use, is mottled with varied and elegant white figures, which are sections of fungiæ, turbinoliæ, caryophilliæ, and other corals, transmuted into opaque calcareous spar. Some of the markings appear like the nebulæ of comets, while others resemble lace-work, being sections of the reticulated structure of the cells of the polypi. I must not, however, dwell longer on this subject, but proceed to the examination of another class of animal existence, which, although not intimately related to the polyparia, it will be convenient to notice in this place, since their remains are found in immense numbers associated with those of corals, in the formations hereafter to be investigated.

37. THE CRINOIDEA, or LILY-SHAPED ANIMALS.—These animals, whose fossilized remains,

* One of the most beautiful examples of this kind I ever beheld is a section of a caryophillia from the Devonshire coast, in the possession of Mrs. Allnutt, of Clapham Common.

cemented together by carbonate of lime, form the
Derbyshire encrinital marble and other beds of
limestone, belong to that division of the animal
kingdom called radiaria, from the different parts
of which they are composed being arranged sym-
metrically around one common centre, as in the
asterias, or star-fish. As the star-fish is so common on our shores, I shall offer a few remarks on
its structure, to illustrate my subsequent observations. In the actinia (page 480), there is no trace
of a skeleton; its tough skin and elastic gelatinous
mass have no support. In the fungia, the polypus
had an immoveable calcareous frame (page 482);
while in the alcyonium, the rudiments of a skeleton are seen in the numerous separate bones, or
spines, dispersed throughout the body: but the
star-fish has an articulated or jointed frame-work
to give stability to the soft parts. The body of the
animal is covered by a tough integument, and each
ray or arm is composed of a series of little bones,
or ossicula, which are united together like the
vertebræ of the spine, and form a flexible yet
powerful support. These bones are often found
in a fossil state, and my museum contains some
interesting examples of fossil skeletons of asterias.
There are some kinds of star-fish which, instead of
the five flat rays of the common species, have
jointed arms, which surround the body and mouth,
like the tentacula of the polypus. These arms are
composed of thousands of little bones, and the

whole are inclosed in the common integument or
skin. The asterias is a free animal; it floats at
liberty in the water. Now, if we imagine a star-
fish, like that which I have described, to possess a
long flexible column, the base of which is attached
to the rock, we shall have a correct idea of the
general character of the Crinoidea, or lily-shaped
animals; which are so called from their fancied
resemblance, when in a state of repose, to a closed
lily.

The name of *Encrinite* is given to the species in
which the bones of the column are circular or
elliptical—that of *Pentacrinite* to those which have
an angular or pentagonal stem. This class of animals contains numerous genera in a fossil state; but
only one living species is known.

38. STRUCTURE OF THE SKELETON OF THE
CRINOIDEA.—The skeleton of the Crinoidea is
formed of numerous ossicula, and consists of an
upright articulated column, permanently attached
by the base, and terminating at the summit in a
cup-like receptacle, or basin, also composed of
bony, jointed plates.

This basin, or pelvis, which originally contained
the body of the animal, is formed of several plates, and
surrounded by long, jointed arms, or tentacula; it
is affixed to the stem by a pentagonal plate placed
in the centre of the base. The column, in most
species, is of a great length, and consists of separate
bones,   articulated,   and   regularly   pierced   in   the

centre, having the articulating surfaces ornamented with radiated, stellular, or floriform markings; the inferior part has a pedicle, or process of attachment, by which the animal was fixed to other bodies. In

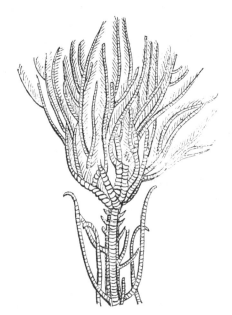

TAB. 59.—SKELETON OF A PENTACRINUS.

(*From the West Indian Seas.*)

the recent state the skeleton was clothed with a fleshy or coriaceous integument; the central perforation in the bones of the column was probably

filled by a medullary, or nervous chord. The crinoidea, like the corals, were permanently attached to
the spot where they grew. The mouth was placed in
the centre of the upper part of the cup, or pelvis, and
the arms, by expanding like the tentacula of the
polypus, could seize their prey, contract and bring
it to the mouth. The only recent species known,
belongs to the genus Pentacrinus (Tab. 59), of
which five or six specimens have been brought to
Europe; it is a native of the West Indian seas.
The number of bones in each skeleton is computed
to exceed 30,000.

The detached ossicula of the crinoidea occur in
myriads in the mountain limestone and transition
rocks; and portions of the stems and separate bones
of one species alone, form extensive beds of limestone in Derbyshire (page 512).

39. THE LILY ENCRINITE.—One of the most
elegant of the fossil crinoidea, is the *Lily Encrinite*
(Tab. 52), which, as I have already stated, occurs
only in the muschelkalk of the New red sandstone
group (page 419), and is principally found in one
locality, near the village of Erkerode, in Brunswick.
The appearance of this zoophyte may readily be conceived from the beautiful specimen before us (Tab.
52), which was formerly in the collection of Mr.
Parkinson. The stem of this species is remarkable,
from being constructed of vertebræ alternately large
and orbicular, and small and cylindrical, thus forming a column of great flexibility. The pelvis of the

Lily Encrinite has the shape of a depressed vase, and it is supposed that the upper part of the cavity was closed by an integument protected by numerous

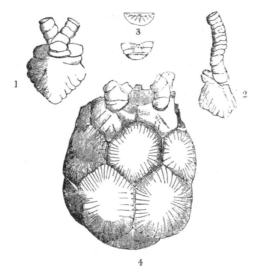

TAB. 60.—MARSUPITE FROM ARUNDEL.

(*Drawn and presented by G. A. Coombe, Esq.*)

Fig. 1. *A plate with one of the arms or tentacula attached.* 2. *Lateral view of one of the arms.* 3. *The ossicula by which the arms are attached to the body.* 4. *The marsupite, with the first five brachial ossicula of two of the arms attached.*

plates, the mouth of the animal occupying the centre. It will elucidate this subject if we examine this specimen of a marsupite, in which the bases of two

of the arms, or tentacula, are preserved (Tab. 60).
A vertebral column attached to the central plate,
at the base of the marsupite, would convert it into
an encrinite; and in the large expanded plates of the
pelvis, and the strong and simple phalanges of the
arms, we have the elements of the more complicated and highly ornamented fabric of the lily
encrinite. In another specimen, the plates which
protected the upper cavity of the pelvis remain.*
The marsupite may be considered as a free encrinus—a link that unites the crinoidea with the star-
fish. The form of the perfect skeleton was shown
in a previous lecture, Tab. 30, page 284.

40. PEAR ENCRINITE OF BRADFORD.—Another
species of crinoidea is known by the name of the
Pear Encrinite of Bradford, the oolitic limestone
quarries in the vicinity of that town abounding in
its remains. The body, or pelvis, of this species
is of a pyriform shape, as its name implies; the
stem is short, smooth, and strong; the arms
are simple, and bear considerable resemblance to
those of the marsupite. In this drawing, from
Mr. Miller's valuable work on the crinoidea, the
animals are represented as if alive in the water
((Tab. 61, fig. 1). This specimen (fig. 2) shows
the state in which the body of the pear encrinite is
usually found; and this vertical section (fig. 3)
the form and disposition of the plates, or bones,

---

* Fossils of the South Downs, Plate XVI. fig. 6.

of which it.is composed.   In the beautiful collection of Mr. Pierce, of Bradford, are specimens of

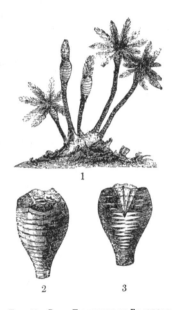

TAB. 61.—PEAR ENCRINITE OF BRADFORD.

(*Apiocrinites rotundus of Miller.*)

Fig. 1. *A group of Apiocrinites restored, and represented as alive in the
water; some with the tentacula expanded, others contracted.* 2. *Body
of the Bradford Encrinite.* 3. *A vertical section of the same.*

the skeleton almost perfect from the base to the
extremity of the tentacula.   In the chalk a small

species of this genus occurs, and its detached ossicula are often met with, impacted in the flints.*

The Pentacrinites differ considerably in their form and structure from the encrinites. The stems are furnished with numerous side arms, as may be seen in the column of the recent specimen (Tab. 59), and the arms sub-divide into innumerable branches. The lias shale of Lyme Regis abounds in the remains of these animals; and large slabs often have the whole surface covered with the plumose tentacula of pentacrinites converted into pyrites. Dr. Buckland has so fully explained the mechanism of these animals, and given so many illustrations, that it is needless for me to pursue the inquiry.† I may, however, observe, that while the bones of the skeleton of a single encrinite may be numbered by tens of thousands, those of the pentacrinite amount to a hundred and fifty thousand; and as each bone must have had its appropriate muscles, the number of the latter, in a single pentacrinus, could not have been less than three hundred thousand.‡

41. DERBYSHIRE ENCRINITAL MARBLE. — I have spoken of limestones formed by crinoideal

---

* Geology of the South-East of England, p. 111.

† Bridgewater Essay, p. 411.

‡ For a more particular account of the natural history of this extraordinary tribe of animals, consult the second volume of Parkinson's Organic Remains; and Miller's Natural History of the Crinoidea, or Lily-shaped Animals, 1 vol. 4to. 1821.

remains. One of the most ornamental of the Derbyshire marbles is almost wholly composed of the

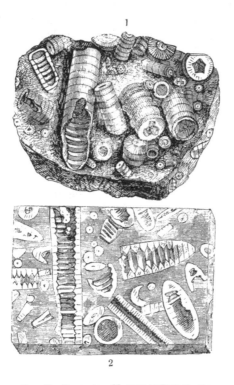

TAB. 62.—ENCRINITAL MARBLE OF DERBYSHIRE.

Fig. 1. *Portions of stems of encrinites, lying in relief on a block of limestone; from Sir George Sitwell, Bart. 2. Polished slab of Derbyshire
marble; from J. Meynell, Esq.*

L L

stems and detached ossicula of one species of encrinite. You are now so familiar with the process by
which the durable remains of animals have been
converted into stone, that it is not necessary to
explain the mode by which this marble has been
formed. In this specimen (Tab. 62. fig. 1), the
encrinital remains are lying in relief ; in the polished
slabs of the Derbyshire marble (fig. 2), so much
used for sideboards, tables, and ornaments, the
sections of the inclosed stems and detached ossicula,
constitute the peculiar and elegant markings of the
limestone. In the chert which occurs in the Derbyshire strata, beautiful casts of the interior of the
columns are met with, the calcareous matter of the
original having been removed ; in this state, the sharp
impressions of encrinital stems form solid silicious
cylinders, deeply marked with annular risings
and depressions, and are called pulley or screw-
stones.*

42. GEOLOGICAL DISTRIBUTION OF THE CRI-
NOIDEA.—One recent species of Pentacrinus is the
only living representative of the once numerous
family of the crinoidea. In the tertiary formations
two species are known ; these have been discovered
in the London clay, by N. T. Wetherell, Esq. of
Highgate. The Chalk contains the marsupite, the
elliptical apiocrinite, and one or more species of pentacrinus. I have also a portion of a crinoideal column

* Organic Remains, Vol. II. plate xv. fig. 6.

that is quadrangular. The Oolite and Lias are more prolific in the remains of this family, yielding about thirty species, belonging to nine genera. The Saliferous system has fourteen species, including the celebrated lily encrinite. The Mountain limestone, and the Slate formation, are the grand repositories of these remains; upwards of fifty species, belonging to twenty-five genera, have been collected from those ancient deposites. The preponderance of these forms of existence in the ocean of those remote epochs, and their almost entire exclusion from the tertiary and modern seas, seem to indicate some remarkable change in the physical condition of the waters during these later periods, inimical to that great development of the crinoidea which existed in the oceans of the transition era.

43. CONCLUDING REMARKS.—From this review of the organization and economy of the polyparia, we learn that an atom of living jelly, floating in the ocean, and at length becoming affixed to a rock, may be the first link in a chain of events, which, after the lapse of ages, may produce important modifications in the physical geography of our globe. We have seen that the living polypi in their rocky habitations enjoy all the blessings of existence, and at the same time are the unconscious instruments of stupendous operations, which in after ages may determine the destinies of mighty nations; and that the materials of their dwellings, consolidated by chemical and mechanical agency, may become the

foundations of islands and continents, and thus
constitute new and favourable sites for the abode
of the human race. When we bring the knowledge
thus acquired to bear on the natural records of our
planet, and examine the rocks and mountains around
us, we find that in periods so remote as to exceed
our powers of calculation, similar effects were produced by beings of the same type of organization
as those whose labours have been the subject of
our contemplation. We are thus enabled to read
the history of the past, and to trace the succession
of events, each of such duration as to defy all
attempts to determine with any approach to probability the period required for its development.

In fine, these investigations have shown us the
marvellous structure of creatures invisible to the
naked eye, their modes of life and action, and the
important physical changes effected by such apparently inadequate agents. They have instructed us
that beings are called into existence, so minute as
to elude our unassisted vision, yet possessing sensation and voluntary motion, each furnished with
its system of muscles and vessels, and preying upon
creatures still more minute, and of which millions
might be contained in a drop of water; nay, even
that these last are supported by living atoms still
less, and so on—and on—till the mind is lost in
astonishment, and can pursue the subject no farther!
Thus are we taught,—

" That those living things
To whom the fragile blade of grass,
    That springeth in the morn
    And perisheth ere noon,
Is an unbounded world,—
    That those viewless beings,
Whose mansion is the smallest particle
Of the impassive atmosphere,
        Enjoy and live like man !
        And the minutest throb,
Which through their frame diffuses
    The slightest, faintest motion,
    Is fixed, and indispensable,
    As the majestic laws
That rule yon rolling orbs !"
                                        SHELLEY.

We have contemplated the results produced by
these countless myriads of animated forms,—the
excess of calcareous matter brought into the waters
of the ocean consolidated by their influence, and
giving birth to new regions; and we have obtained
evidence that in the earlier ages of our globe like
effects were produced by similar living instruments.
The beds of fossil coral are now the sites of towns
and cities, whose inhabitants construct their abodes
of the limestone, and ornament their temples and
palaces with the marble, formed of the petrified
skeletons of the zoophytes, which lived and died in
oceans that have long since passed away !

Hence we perceive that He who formed the
Universe creates nothing in vain ; that His works all
harmonize to blessings unbounded by the mightiest
or the most minute of His creatures ; and that

the more our knowledge is increased, and our
powers of observation are enlarged, the more exalted will be our conception of His wondrous
works.   Thus, in the eloquent language of Dr.
Chalmers,—" while the telescope enables us to see
a system in every star, the microscope unfolds to
us a world in every atom.   The one instructs us
that this mighty globe, with the whole burden of
its people and its countries, is but a grain of sand
in the vast field of immensity—the other, that
every atom may harbour the tribes and families of
a busy population.   The one shows us the insignificance of the world we inhabit—the other re-
deems it from all its insignificance, for it tells us
that in the leaves of every forest, in the flowers of
every garden, in the waters of every rivulet, there
are worlds teeming with life, and numberless as are
the stars of the firmament.   The one suggests to
us, that above and beyond all that is visible to man,
there may be regions of creation which sweep im-
measurably along, and carry the impress of the
Almighty's hand to the remotest scenes of the
Universe—the other, that within and beneath all
that minuteness which the aided eye of man has
been able to explore, there may be a world of invisible beings ; and that could we draw aside the
mysterious curtain which shrouds it from our senses,
we might behold a theatre of as many wonders as
astronomy can unfold ; a Universe within the com-
pass of a point, so small as to elude all the powers

of the microscope, but where the Almighty Ruler of
all things finds room for the exercise of his attri-
butes, where he can raise another mechanism of
worlds, and fill and animate them all with the evidence of his glory !" *

* This Lecture was illustrated by a splendid collection of
recent and fossil corals ; and specimens of living actiniæ and
zoophytes, from the neighbouring sea.   Among the former
were fine examples of caryophilliæ and meandrinæ, contributed
by Mrs. Robertson; fungiæ, by Miss Crofts; madreporæ, den-
drophylliæ, and astreæ, by Lady Mantell ; agariciæ, by Miss
Ellen Gladwin; gorgoniæ and milleporæ, by William Tennant,
Esq.; flustra foliacea, by Robert Hannay, Esq.; and of turbinoliæ and milleporæ, by Thomas Bodley, Esq.— G. F. R.

LECTURE VII.

1. INTRODUCTION.—The examination of the recent and fossil zoophytes, which formed the subject of the last discourse, will enable us to comprehend many of the phenomena relating to the ancient coralline rocks hereafter to be noticed. I now resume the geological argument from which we have for a while digressed, and hasten to the consideration of the Carboniferous system, which in the chronological arrangement (page 179, Pl. 3,

fig. 7) succeeds the saliferous deposites described in the Fifth Lecture.

The strata comprised in the Carboniferous system, so named from its being the great depositary of that important substance called coal, admit of three natural divisions. The uppermost is composed of a vast number of alternations of coal, shale, ironstone, and sandstone; the middle, of limestone, chert, and sandstone, with immense quantities of shells, polyparia, crinoidea, and other marine exuviæ; and the lowermost, of sandstones and conglomerates, generally of a dull red colour, and bearing some resemblance, in their lithological characters, to those of the new or upper red sandstone. I propose to describe—firstly, the general features of the strata, and their geographical distribution; secondly, the nature of the coal and of the fossil plants, which are scattered through the carboniferous rocks; thirdly, to describe the animal remains; and lastly, to review the flora of the ancient world.

2. THE CARBONIFEROUS SYSTEM.—The following tabular arrangement will convey a general idea of the characters and relations of these deposites.

1. THE COAL MEASURES.

(*The uppermost in the series.*)

Sandstone, shale, and immense beds of coal; with layers of ironstone irregularly stratified, abounding in terrestrial plants.

Beds of limestone, with fresh-water shells.

Total thickness, 1000 yards.

2. THE CARBONIFEROUS, OR MOUNTAIN LIMESTONE.

Millstone grit, sandstone, shale, and coal, with plants.

Limestone and flagstone, abounding in crinoidea, with plants.

Lower, or scar-limestone, with zoophytes in profusion, crinoi-
    dea, productæ, spiriferæ, orthoceratites, ammonites, trilo-
    bites, &c.

Total thickness, about 800 yards.

3. OLD RED SANDSTONE.

Conglomerates and silicious sandstones, without organic re-
    mains.

Flagstones, marls, and concretionary limestones; scales of
    fishes, orthoceratites and nautili occur, but organic remains
    are very rare.

Total thickness, 3500 yards.

Such is a synoptical view of the strata compre-
hended in this series, which I shall now proceed to
examine somewhat in detail.

3. THE COAL MEASURES. — The bituminous
substance termed coal is simply vegetable matter
altered by a chemical process, which will hereafter
be explained. It occurs in strata which vary from
a few inches to a fathom in thickness, having interposed layers of shale, clay, sandstone more or less
micaceous, and ironstone in layers and nodules.
Groups of alternations of this kind occupy circumscribed areas, or basins, as they are termed. Mr.
Bakewell observes that the strata thus disposed may
be explained by a series of muscle shells, placed
one within the other, and having layers of clay
interposed. If one side of the shell be raised, to

indicate the general rise of the strata in that direction, and the whole series be dislocated by partial cracks or fissures, the general arrangement of the beds and the displacements will be represented; each shell being the type of a bed of coal, and the partitions of clay of the earthy strata which separate the carboniferous layers.

The usual characters of the strata of a coal field, as a series of the kind is termed, are shown by this section of the South Gloucestershire coal basin (Pl. 5, fig. iv.), from Mr. Conybeare.* Here you perceive that the old red sandstone, the lowermost group of the carboniferous system, has been elevated almost into a vertical position, and a layer of mountain limestone lies immediately upon it, and partakes of the same inclination. This is succeeded by a corresponding bed of millstone-grit, which is followed by alternations of coal and grit. The unconformable position of the lias and inferior oolite is here shown (see page 409). The mountain limestone and millstone grit are seen on the opposite flank (*on the left*) of the elevated ridge of old red sandstone. It will be instructive to enumerate the deposites exhibited in this section, in their chronological order; that is, in their relative position if they were piled upon each other, and had suffered no displacement: commencing with the lowermost or most ancient.

* England and Wales, p. 428.

— Old red sandstone (Pl. 5, fig. iv.) of the Mendip Hills.
1. Mountain limestone.
2. Millstone grit.
Alternations of coal and shale, and pennant grit.
3. New red sandstone.
4. Lias.
5. Inferior oolite.
6. Great oolite.
7. Oxford clay, south of Malmesbury.

In another line of section in the same district, we perceive the same general arrangement.

Old red sandstone.   Pl. 5, fig. ii. *left hand section.*
Carboniferous limestone.
New red sandstone.
Lias.
Inferior oolite.

The term basin, applied to these accumulations of coal, must be taken only in a general sense; for although some carboniferous deposites may have been found in circumscribed depressions, it appears probable that the beds have extended over large areas; and that their present isolated and confined limits may be attributable to subsequent elevations of the rocks on which they repose, and by which the faults and dislocations of the coal-beds have been produced.

4. COAL-FIELD OF DERBYSHIRE.—The Derbyshire coal-field, so admirably illustrated by my friend Mr. Bakewell,* will serve as a type of the English

* Introduction to Geology, p. 149.

series. The geographical distribution of the principal coal-measures in England is shown in this
map (Pl. 6). The strata of limestone, which form
the grand mountain-chains of Derbyshire, decline
towards the eastern side of the county, and sink
beneath the coal-measures. Immediately upon the
carboniferous limestone, is placed a bed of calcareous slate or shale, varying in thickness from
three to six hundred feet. The compact strata
of this bed are separated by softer layers, which
readily disintegrate ; and they form the exposed
face of Mam-Tor, or the shivering mountain near
Castleton. These are succeeded by a mass of grit,
or conglomerate, with vegetable remains, which
is worked for mill-stones ; hence the geological
name by which it is distinguished. Above the grit
are the regular coal strata, comprising sandstones
of various qualities and frequently in exceedingly
thin laminæ ; indurated clay ; ironstones, the nodules
of which contain organic remains; and softer argillaceous beds, which being of a slaty structure, are
therefore called shale. Two of these layers of clay
abound in fresh-water muscle shells, of extinct
species, and are termed *muscle-bind ;* these bivalves
very much resemble the uniones of the Wealden
(page 344). The total thickness is 1310 yards,
which includes thirty different beds of coal, varying
from six inches to eleven feet, and making the
total thickness of the coal about twenty-six yards.
In the beds of limestone shale below the coal strata,

there is a transition from marine calcareous strata, with animal remains, to fresh-water deposites, with terrestrial vegetables : this may have originated from occasional intrusions of freshes from a river.

The series above enumerated is often repeated ; similar shales, clays, and sandstones occurring under different beds of coal, with a perfect similarity both in the succession and thickness of each. Interruptions to the continuity of the strata, from cracks or fissures which have taken place since their original deposition, are very frequent. In this diagram (Pl. 3, 7), a fault of this kind (see page 177) is represented as having displaced three beds of coal. This is an example of a simple fissure and dislocation ; but dykes, or intrusions of other mineral matter, also occur, separating the beds as it were by vertical walls, which are from a few inches to many yards in thickness. The dykes are sometimes composed of indurated clay, but more frequently of the ancient volcanic rock, called trap, or basalt. This diagram (Pl. 3, 15) represents a trap dyke intruding through the carboniferous and other systems, and spreading over the chalk. The effect of these lava currents on the rocks they traverse, will be considered in the next lecture. The mottled rock, called *toad-stone* in Derbyshire, is clearly a pyrogenous rock, that has been erupted in a melted state ; and in Yorkshire the limestone strata are traversed by a dyke of basalt and greenstone, called *whin-sill* (see the map, Pl. 6).

5. COALBROOK DALE. — Coalbrook Dale in Shropshire is situated on the east side of the range of transition rocks forming the Wrekin and Wenlock Edge, and the coal strata are superposed on millstone grit. Beds of ironstone occur, abounding in nodules, with organic remains, of which I shall speak hereafter. This coal-field is remarkable for the dislocated and shattered state of the strata, and the intrusion of volcanic rocks, which do not appear as dykes, or in the fissures of the beds, but rise up in mounds or protuberances. The walls of the fissures are in some instances several yards apart, the intervals being filled with debris. Beds containing marine shells, alternate with others abounding in fresh-water shells and land plants, as in Derbyshire.* The series of strata forming this carboniferous accumulation, consists of quartzose sandstone, indurated clay, slate-clay, and coal. A pit sunk in Madely colliery, in a depth of 730 feet, passed through eighty-six beds of alternating quartzose sandstone, clay-porphyry, coal, and indurated clay, containing argillaceous ironstone in nodules. The sandstones of Coalbrook Dale are fine-grained and micaceous, and some beds are penetrated by *petroleum*, which at Coalport escapes from the surface in a tar spring ; bitumen also occurs in some of the shales. Plants, shells, and crustacea

* See a highly interesting memoir on this coal-field by Mr. Prestwich.

are abundant in the shale and ironstone nodules; and the remains of insects sometimes occur.

This brief notice of two remarkable coal-fields will suffice to convey a general idea of the nature of carboniferous deposites. To exemplify those of the United Kingdom alone would require a course of lectures. The admirable memoirs on our coal-fields in the Geological Transactions, by some of our most eminent observers, and in the works of Bakewell, Conybeare, Phillips, Lyell, De la Beche, Dr. Buckland, and others, will afford those who wish to pursue the inquiry, information of the most important and interesting nature.

6. COAL-SHALES AND VEGETABLE REMAINS.— The layers of shales, or slaty-coal, are the great depositories of the fossil vegetables. These strata intervene between the beds of bituminous coal, and when the latter are exhausted, the roof and floor of the mines or galleries are composed of the schistose beds, which are not made use of for economical purposes. In these large slabs of shale, from New-castle, for which I am indebted to William Hutton, Esq., you perceive that vegetable remains occur between every lamina, and as I flake off these portions bearing the leaves of ferns and other plants, another series is disclosed ; the whole mass being formed of leaves and stems, pressed together in clay. The roof of a coal-mine when newly exposed displays the most interesting appearance of this kind imaginable; leaves, branches, and stems of

the most elegant and delicate forms, being embossed on the dark shining surface.

The coal-mines of Bohemia, the fossil plants of which are well known, from the beautiful work of Count Sternberg, are stated by Dr. Buckland to be the most interesting of any he had visited; but I will describe them in his own eloquent language:—
" The most elaborate imitations of living foliage on the painted ceilings of Italian palaces, bear no comparison with the beauteous profusion of extinct vegetable forms, with which the galleries of these instructive coal-mines are overhung. The roof is covered as with a canopy of gorgeous tapestry, enriched with festoons of most graceful foliage, flung in wild irregular profusion over every portion of its surface. The effect is heightened by the contrast of the coal-black colour of these vegetables, with the light ground-work of the rock to which they are attached. The spectator feels transported, as if by enchantment, into the forests of another world; he beholds trees of form and character now unknown upon the surface of the earth, presented to his senses almost in the beauty and vigour of their primeval life; their scaly stems and bending branches with their delicate apparatus of foliage are all spread forth before him, little impaired by the lapse of countless ages, and bearing faithful records of extinct systems of vegetation, which began and terminated in times of which these relics are the infallible historians. Such are the grand

M M

natural herbaria wherein these most ancient remains
of the vegetable kingdom are preserved in a state of
integrity, little short of their living perfection, under
conditions of our planet which exist no more."*

7. CARBONIFEROUS, or MOUNTAIN LIMESTONE.
—The deposites comprised in this geological group,
consist, 1st, of Millstone-grit :—2, Bluish-grey limestone traversed by veins of calcareous spar, abounding
in encrinital remains and other marine exuviæ; in
some localities it is rich in lead ore ; hence it has
been called metalliferous limestone :—3, Chert, sandstone, shale, and coal of inferior quality.

The millstone-grit is a silicious conglomerate
composed of the detritus of primary rocks.   In
these specimens, rolled fragments of granite of
various magnitudes, from the size of a pea to that
of a large pebble, are cemented together by a crystalline paste.   Silicious sandstones are associated
with the grit, and have one common origin : the
materials of which they are composed are clearly
the detritus of granitic rocks, produced by the
action of water.   Fragments of shale, coal, red
sandstone, stems of plants, &c. are occasionally
found imbedded in the grit.   The millstone grit, however, is but of limited extent, and in some localities
is altogether wanting, or is superseded by chert, or
sandy limestones.

The carboniferous or mountain limestone is of a
sub-crystalline texture, and many varieties are suf-

* Bridgewater Essay, p. 458.

ficiently compact to bear a fine polish, their surfaces being ornamented by the sections of inclosed crinoidea, corals, and shells. The prevailing colour is a light bluish-grey, the organic remains being of a pure white ; but some varieties have a ground of pale red, while others are nearly black, the imbedded shells having a deep ochreous colour. The Derbyshire marbles, and those of St. Vincent's rock, at Clifton, are familiar examples of the finer varieties of the mountain limestone. It is largely developed over the central and northern parts of England, and south-west of Scotland ; and is the predominating rock throughout a great part of Ireland. In Somersetshire, Gloucestershire, Shropshire, North and South Wales, and Derbyshire, this limestone constitutes as it were an entire calcareous mass, interposed between the old red sandstone, or when that is wanting, the more ancient slate rocks below, and the sandstone and shales of the coal above. In Cumberland and Westmoreland, &c. it forms an elevated belt, which partly surrounds the Cumbrian slate mountains, and forms, on the west, a ridge nearly three thousand feet in height. In Derbyshire it constitutes the grand physical feature of the country, rising into crags or peaks, and hills, which present bold escarpments, and exhibit the wildest and most picturesque scenery. Professor Phillips estimates the thickness of the lower division of limestones with shale partings (provincially termed scar-limestones), in Derbyshire, at

M M 2

750 feet; the alternations of shale, sandstone, lime-stone, and ironstone, which surmount the former, at 500 feet; and the cappings of millstone-grit which form the summits of the hills, at 360 feet.

The encrinital or Derbyshire marble (page 513,) has already been described; marbles with shells and corals also occur, and are employed for ornamental purposes. There are three intrusions or beds of *toad-stone*, a volcanic rock, in the mountain lime-stone of Derbyshire. In some parts of this district, and in the Mendip hills, layers and nodules of the silicious substance called chert, occur imbedded in the calcareous rock, like the flints in chalk. The masses of chert before us, collected by Sir George Sitwell, are fine examples of the curious fossils called screw-stones, of which I spoke in the last lecture (page 514). The chert or silex has flowed into the interior of the stems of the encrinites, and the cavities of the shells, and become consoli-dated; the calcareous part of the organic bodies has since decomposed, leaving the chert, which ex-hibits beautiful casts of the interior, and impressions of the external surface. So abundant are organic remains in some beds of the mountain limestone, that it is computed corals, shells, and crinoidea constitute at least three-fourths of the mass.

8. DERBYSHIRE LEAD-MINES.—It is in the mountain limestone that the principal British lead-mines are situated, namely, those of Somersetshire, Derbyshire, York, Durham, and Northumberland.

In Derbyshire the metal occurs in numerous veins which traverse the rock, and extend in some instances into the *toad-stone*, the ancient volcanic bed, which I have already mentioned. The perpendicular, or rake-veins as they are termed, are from two to forty feet wide; and there are chasms or hollows in the rock, several hundred feet in width, which also contain metallic ores and spar. Manganese, copper, zinc, and iron are found in the limestone, but the predominating metalliferous ore, is the sulphuret of lead, or galena, as it is called by mineralogists. This substance, as you perceive in the specimens before you, is of a bluish-grey colour, and sometimes occurs in cubic and octahedral crystals; it is also found disposed in thin layers, as well as in veins. It is accompanied by fluor and calcareous spar, sulphate and carbonate of barytes, iron pyrites, &c. The variety termed specular galena, or *slickensides*, is a thin coating of lead on the sides of the veins, and appears to have arisen from one wall of the fissure having slipped along the face of the other, so as to give it a polished or *slicken* surface.

The beautiful substance known by the name of Derbyshire spar, is fluate of lime, and occurs in crystals, and also in nodular masses. In the Odin mines, near Castleton, fluor is found in veins, and in masses from three inches to a foot in thickness; the celebrated *blue-john*, of which elegant vases and other ornaments are made, is found in this state.

9. The Old Red Sandstone.—The old red sandstone, or red marl, lies beneath the mountain limestone, and is largely developed in Herefordshire, Monmouthshire, and in the south-east border of the Grampians.   It consists of many varieties and alternations of conglomerates, shales, and sandstones.   The sandstones are in various states of induration, and when schistose, are employed for roofing.   The conglomerates contain abundance of quartz pebbles.   The red colour predominates in the cementing material and in the marls, and is derived, like that of the New red, from the peroxide of iron.   The formation of this group of strata has manifestly resulted from the waste and degradation of the ancient slate rocks, of which I shall hereafter speak; the detritus being cemented together, more or less compactly, by red sand or marl, into coarse conglomerates.   The slate rocks of Cambria, elevated by convulsive movements, were subjected to degradation; and accumulations of pebbles, sand, and mud, collected in hollows or depressions of the sea.   The mountains of Scotland are bordered by enormous depositions of a like character; and those of North and South Wales by extensive beds of pebbly red sandstone.* It would appear from the observations of some of our most eminent geologists, that prior to the formation of the old red sandstone, the Cambrian slaty group of sands, rocks, flags, &c. with porphy-

* Phillips.

ritic conglomerates, had been long consolidated;
and that subterraneous movements elevated them,
threw the strata on edge, and formed an irregular
island: at the same time, the Grampians, Lammer-
muirs, and the slaty tracts of Ireland and Wales,
and the Ocrynian chains of Cornwall, stood above
the waters. As a whole, this division of the carboniferous system contains but few organic remains;
but, locally, certain fossils abound. These consist of
the ancient forms of the family of the terebratulæ,
the spiriferæ, productæ, &c.; nautili, ammonites,
and the orthoceratite, a singular concamerated shell,
which I shall hereafter describe. Fishes of a peculiar character occur, and scales of some unknown
species. Fuci are, I believe, the only known vegetable remains.

10. GEOGRAPHICAL DISTRIBUTION OF THE
CARBONIFEROUS SYSTEM.—I have briefly alluded
to the distribution of the carboniferous system of
England,* and further detail would be irrelevant to
the plan of these lectures. I cannot, however, omit
to notice one most interesting locality, from which
many of the specimens before us were collected.

The magnificent gorge of the Avon at Clifton,
so well known by the name of St. Vincent's Rocks,

* The principal coal basins in England are those of Somer-
setshire, Gloucestershire, North and South Wales, Dudley,
Shropshire, Leicestershire, Lancashire, Nottingham, Derby-
shire, Yorkshire, Cumberland, Durham, Newcastle; of the
Forth and Clyde; and the central districts of Ireland.—See
Phillips's Guide to Geology.

is lined by an uninterrupted succession of mural precipices, and affords an unrivalled natural section of the carboniferous limestones. The calcareous beds rest on conformable strata of old red sandstone, which may be seen on both sides the river, near Cook's Folly, extending on the south under Leigh-down and Weston-down. On the north the old red sandstone passes towards Westbury.*

On the continent, coal, with limestones and red conglomerates, in some instances resembling, in others differing from, the English strata, are known in France, near Boulogne, Mons, and St. Etienne; in the Low Countries, at Namur and Liege; in Germany, Silesia, Moravia, Poland, and in the Carpathian Mountains. The mountain limestone tract along the Meuse, in the Netherlands, resembles that of Derbyshire and Monmouthshire, and appears to be of the same age; and the scenery to which it gives rise reminds the English traveller of the banks of the Derwent or the Wye.†

The coal deposites of central France repose on granite and other primary rocks, without the intervention of limestones or sandstones.‡ In Poland, the lower beds of the coal measures pass insensibly into the transition rocks upon which they rest. In North America, the carboniferous series is largely developed, and has been ably illustrated by Pro-

---

* From Mr. Conybeare's admirable description of the carboniferous system of Somersetshire, &c.

† Phillips.   ‡ De la Beche.

fessor Silliman, Eaton, Hitchcock, and other American geologists. The *coal* of the United States
appears, however, to be referrible to different geological eras; the most ancient belonging to the
transition series—the next to the European carboniferous group—and the third to the brown coal, or
tertiary lignite. The coal, or anthracite, of Pennsylvania, of which, through the kindness of Professor Silliman and Dr. Morton, I have a fine
series, with vegetable remains, is associated with
conglomerates, sandstones, and argillaceous slate ;
and the conglomerate is composed of quartz pebbles, like our old red sandstone. In the valley of
the Connecticut, bituminous coal is stated by Professor Hitchcock to be intercalated in a group of
strata, which he refers to the new red sandstone.*
Deposites of anthracite, or stone coal, exist in
Worcester and in Rhode Island, of which an admirable account has been published by Professor
Silliman. Extensive coal fields occur to the west
of the Alleghany Mountains, towards the Mississippi. The base of the whole extent of the plain of
that mighty river appears to be mountain limestone;
it has been perforated to a depth of six hundred
feet, and contains trilobites, orthoceratites, and other
remains, which I shall hereafter point out to you as
characteristic of this formation. The limestone
extends under the Alleghany Mountains in the
east, and the sand plains on the west, and rests on

* Geology of Massachusetts.

the granite rocks of Canada on the north. The uppermost layer of the mountain limestone supports strata of bituminous coal and shale. This coal-field is 1500 miles in length, and 600 in breadth. Ironstone abounds in these deposits, and mines of lead occur over a district of two hundred square miles, between the Missouri and the Illinois. Thus the coal basin of the Mississippi appears to possess all the essential features of the English carboniferous series.*

Coal is found in Asia; and has long been worked in China. In Van Diemen's Land, carboniferous strata occur, associated with sandstones, and yield coal abundantly.†

11. VOLCANIC ROCKS OF THE CARBONIFEROUS EPOCH.—I have alluded to the intrusions of volcanic matter, which are found in the carboniferous strata, and I will now offer a few remarks on the phenomena which these pyrogenous rocks exhibit. One of the most remarkable and well-known volcanic substance of this era is the rock called *toadstone*, so named from its being mottled with green and yellow. Three distinct beds of this ancient lava are interpolated in the mountain limestone of Derbyshire.‡ The toadstone is a compact mass, consisting of small nodules of white and yellow

---

* Stuart's Travels in the United States.

† Conybeare and Phillips.

‡ Consult Mr. Bakewell's Geology, and Messrs. Conybeare and Phillips's Geology of England and Wales.

calcareous spar and of green earth, imbedded in a dark greenish paste of basalt. Sometimes the nodules are decomposed, and the stone then is vesicular or cellular, resembling porous lava. This substance, in some instances, passes into common basalt, an ancient volcanic product, which I shall describe in the next discourse. The thickness of each of the beds of toadstone varies from sixty to eighty feet. In some instances, the dykes of this substance traverse the metalliferous veins, and a manifest alteration is then observable in the nature of the latter.

Trap dykes, which are intrusions of a hard, dark green, fine-grained, volcanic stone, in fissures which intersect the stratified deposites, are often seen in the carboniferous series; and there is one of prodigious extent and thickness, known by the name of the *Whin sill*, which traverses the coal-measures, red sandstone, and lias, and passes from High Teesdale to the confines of the eastern coast, a distance of upwards of sixty miles.* The alterations produced in the strata in contact with these erupted lava currents will be considered in the next lecture ; at present it is only necessary to state, that changes of the most striking character occur in the coal and its associated beds, wherever they have been within the range of the volcanic influence ; the coal being charred, and deprived of its bitu-

* Professor Sedgwick. Mr. Bakewell's observations on the phenomena of trap dykes are highly interesting and philosophical.

minous quality, and sometimes changed into anthracite. I shall have occasion to revert to this subject
hereafter.

12. ORGANIC REMAINS OF THE CARBONIFEROUS
SERIES.—The animals and plants entombed in the
carboniferous group are exceedingly numerous, and
well-defined. In the coal, the vast abundance of
terrestrial plants, the presence of fresh-water shells
with but few marine species, together with the
entire absence of zoophytes, crinoidea, and other
marine exuviæ, which abound in the mountain
limestone, present a remarkable contrast with the
calcareous deposites of this system. The shales of
the limestone, however, contain vegetable remains.
Marine fishes, shells, corals, and crinoidea, abound
in the limestone, but are of rare occurrence in the
old red sandstone.

It is a circumstance worthy of remark, that the
coal beds of the south of England do not contain
marine remains, whereas in the north, certain beds
abound in ammonites, and other inhabitants of the
sea. At Burdie House, near Edinburgh, beds of
limestone, with fresh-water shells and crustacea,
sauroid fishes, and terrestrial plants, have been discovered in the carboniferous series by Dr. Hibbert,
and appear to be an intercalation between the
marine deposites of the same group.*

* On the Freshwater Limestone of Burdie House, near
Edinburgh, by Samuel Hibbert, M.D. F.R.S. Edinburgh.
Philosophical Transactions, vol. xiii.

In the immense accumulations of the early vegetation, of which the coal-measures are composed,
we have presented to us, in the most legible and
striking characters, the peculiar flora of the remote
epoch in which those deposites were produced.
But to obtain any satisfactory results from an examination of these remains, some knowledge of the
internal structure of vegetables is requisite ; for
in a fossil state the mere external characters are,
for the most part, either so imperfect or obliterated,
as to afford but obscure indications of the nature of
the original. And as in our investigations of the
fossil remains of animals, we availed ourselves of
the principles of comparative anatomy (page 114)
to reconstruct those extinct forms of being; in like
manner we must now call to our aid that branch
of science which treats of vegetable organization,
and we shall thus be enabled to restore anew the
forests of extinct palms and tree-ferns, the groves
of liliacea, and all the luxuriant tropical vegetation
which flourished in the carboniferous epoch of our
globe. I must, however, restrict myself to a very
brief enunciation of a few botanical principles. The
works of M. Adolphe Brongniart,* and of Dr.
Lindley and Mr. Hutton,† should be consulted by

* Histoire des Végétaux Fossiles, ou Recherches Botani-
ques et Géologiques, &c. par M. Adolphe Brongniart. 1 vol.
4to. with numerous plates.

† The Fossil Flora of Great Britain, by Dr. Lindley and
W. Hutton, Esq. 8vo. publishing in parts.

those who would pursue this most attractive department of natural history.*

13. ORGANIZATION OF VEGETABLES.—In the
previous discourse, the complex organization of
even the most minute forms of animal existence
was remarked; the structure of vegetables, on
the contrary, presents a remarkable simplicity.
While in animals every separate function requires
an organ of peculiar construction, in plants a few
tissues, variously modified, constitute the mechanism
by which all the vegetable functions are performed.
The section of any living plant shows that the intimate structure consists of a solid spongeous texture,
made up of cells or vessels, containing fluids, or
other matter. This structure is differently arranged
in the grand classes of the vegetable kingdom. In
the most simple group, the cellulares, called also
the acotyledones, from the absence of seed-lobes,
the tissue is wholly cellular, the cells being nearly
of equal size and consistence; mosses, lichens, sea-
weeds, fungi, &c. are examples. These plants have
no flowers, and hence are named *cryptogamic*.
The vegetables belonging to the other great class
are termed *vasculares*, from their cellular tissue
being more complex, and assuming the structure of
tubes and vessels; and *phanerogamic*, from their
bearing flowers. Their tissue is composed of cells
of various sizes and forms, and of straight and spiral

* See also Henslow's Principles of Physiological Botany:
a most instructive and attractive volume.

tubes.   This class is subdivided into two families,
viz. the *Monocotyledonous,* so called from the seed

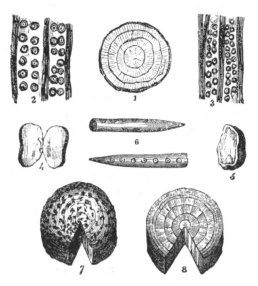

Fig. 1. *Section of a coniferous tree, showing the concentric and radiated
structure.   2. Longitudinal section of a pine tree, magnified to exhibit
the spotted vessels.   3. Longitudinal section of the vessels of an Auracaria, magnified, to show the glands or ducts arranged alternately.   4. A
dicotyledonous seed split open : the germ is in the middle.   5. Section of
a monocotyledonous seed, with the germ below.   6. Dotted tubes, or eells of
coniferæ.   7. Section of a monocotyledonous stem.   8. Section of a
dicotyledonous tree, showing concentric circles, medullary rays, and
the central pith.*

having but one fleshy lobe, or *cotyledon* (Tab. 63,
fig. 5), as the onion, lily, &c. ; and also called *endo-*

genæ (*from within*), because increase takes place
from the innermost part of the stem; and the *dicotyledonous,* from the seeds having two lobes (Tab.
63, fig. 4), as the bean, almond. &c ; these are also
termed *exogenous* (*from without*), the new matter
being added externally to the old layers, and thus
forming annual circles of increase, as in the oak,
elm, &c. (Tab. 63, fig. 8.)   The section of the monocotyledonous stems (as the cane, palm, &c.) present,
therefore, openings of tubes, which are condensed
towards the outer surface (Tab. 63, fig. 7), while the
dicotyledonous exhibit annular lines of growth with
diverging rays, and a central pith (Tab. 63, fig. 8).

14. CONIFEROUS TREES.—In some groups of dicotyledonous trees the elongated cells, or tubes, are
studded with spots or glands (Tab. 63, fig. 6), and this
is particularly the case in the woody fibres of the coniferæ. In this magnified view of a slice of the common
deal (Tab. 63, fig. 2) the spots or glands are seen to
be arranged in double parallel lines. In a remarkable
family of pines, the *Auracaria,* the spots are placed
alternately, and sometimes in triple rows.   The
name of coniferæ (*cone-bearing*) is derived from
the fruit of these plants being in the form of a cone,
as in the fir, larch, &c.; transverse sections of
the stems show the concentric layers and radiated
structure of the dicotyledones.   All the trees of
this order secrete resin, have branched trunks,
and linear, rigid, entire leaves : species are found
in the coldest as well as in the hottest regions.

The *auracaria* I have just mentioned, is a native of Norfolk Island, a small spot in the South Pacific, about fifteen miles in circumference. This island presents a scene of the most luxuriant vegetation, and abounds in a particular species of pine, the auracaria, or *altingia excelsa,* which attains a height of two hundred feet, and a circumference of thirty : it will not thrive in the open air in this country.

My limits are too circumscribed to admit of dwelling at length on any other of the numerous co-relations of structure presented in vegetable organization. I will only add, that even in the foliage of the different orders, there are such evident distinctive characters, that the botanist could, from a mere fragment of a leaf, detect the dicotyledonous structure in the fibrous interlacing of its vessels, as in that of the oak ; and the monocotyledonous in the smooth parallel veins of the lily.* The application of these principles to the investigation of fossil plants I will now consider.

15. CLIMATE, SEASONS, AND PERIODS OF TIME INDICATED BY FOSSIL WOOD.—In the course of these lectures, it has been repeatedly demonstrated, how, by a knowledge of comparative anatomy, the forms, structure, and economy of beings long since obliterated from the face of the earth, may with

* The general reader will find the organization of plants explained in a clear and pleasing manner, in a little volume entitled, " POPULAR BOTANY," by James Main, Esq., A. L. S. London, 1835.

N N

certainty be determined. I now proceed, by the
aid derived from a few botanical principles, to illustrate not only the form and character of vegetables,
of which but the faintest vestiges perhaps remain,
but also to point out the important inferences at
which we may arrive, relating to the state of the
earth, the nature of the climate, and even of the
seasons which prevailed at the periods when those
plants flourished. Our distinguished countryman,
Professor Babbage, has so forcibly exemplified the
inductive process by which such results may be
obtained, that I shall largely avail myself of his
interesting remarks.

" We have seen," observes this distinguished philosopher,
"that dicotyledonous trees increase in size by the deposition of
an additional layer annually between the wood and the bark;
and that a transverse section of such trees presents the appearance of a series of nearly concentric, irregular rings, the number of which indicates the age of the tree. The relative thickness of these annular markings depends on the more or less
flourishing state of the plant during the years in which they
were formed. Each ring may, in some trees, be observed
to be subdivided into others, thus indicating successive periods
of the same year during which its vegetation was advanced or
checked. These rings are disturbed in certain parts by irregu-
larities resulting from branches; and the year in which each
branch first sprang from the parent stock, may therefore be ascertained by proper sections. These prominent effects are
obvious to our senses; but every shower that falls, every
change of temperature that occurs, and every wind that blows,
leaves on the vegetable world the traces of its passage; slight,
indeed, and imperceptible perhaps to us, but not the less permanently recorded in the depths of those woody fabrics.

" All these indications of the growth of the living tree are preserved in the fossil trunk, and with them also frequently the history of its partial decay.   Let us now examine the use we can make of these details relative to individual trees, when considering forests submerged by seas, imbedded in peat mosses, or transformed, as in some of the harder strata, into stone.   Let us imagine that we possessed sections of the trunks of a considerable number of trees, such as those occurring in the bed called the *Dirt-bed* in the Island of Portland (page 235).   If we were to select a number of trees of about the same size, we should probably find many of them to have been contemporaries.   This fact would be rendered probable if we observed, as we doubtless should do, on examining the annual rings, that some of them, conspicuous for their size, occurred at the same distances of years in several trees.   If, for example, we found on several trees a remarkably large annual ring, followed at the distance of seven years by a remarkably thin ring, and this again, after two years, followed by another large ring, we should reasonably infer from these trees, that seven years after a season highly favourable to their growth, there had occurred a season highly unfavourable to them : and that after two more years, another very favourable season had happened, and that all the trees so observed had existed at the same period of time.   The nature of the season, whether hot or cold, wet or dry, would be known with some degree of probability, from the class of tree under examination.   This kind of evidence, though slight at first, receives additional and great confirmation by the discovery of every new ring which supports it ; and, by a considerable concurrence of such observations, the succession of seasons might be ascertained in geological periods, however remote."

16. VERTICAL TREES IN CARBONIFEROUS STRATA.—The occurrence of the trunks of fossil trees in a vertical position, has already been noticed ;

N N 2

the petrified forest of Portland, with its beds of vegetable mould, having early drawn our attention to this phenomenon. Stems of plants, standing upright in the strata, occur in many coal mines; and a very interesting example is described by my friend M. Alexandre Brongniart, a distinguished French philosopher, as occurring at Tréuil, near St. Etienne, in the department of the Loire. This mine is most favourable for observation, for it is, in truth, a quarry in the open air, and exposes a natural section of the strata, which consist of clay-slate and coal; with four layers of compact iron ore, in flattened nodules, which are accompanied, and even penetrated, by vegetable remains. The upper ten feet of the quarry is composed of micaceous sandstone, in some instances stratified, and in others possessing a slaty structure. In this bed are numerous vertical stems traversing all the strata, and appearing like a fossil forest of plants resembling the bamboo, or large equiseta, turned into stone, in the places where they grew. The stems are of two kinds: the one long and slender, from one to four inches in diameter, and nine or ten feet high, being simply jointed and striated solid cylinders of sandstone, with a thin coaly envelopement, or crust. The other, and less common species, are hollow, cylindrical stems, spreading out from the base, like a root, but without ramifications.*

* Notice sur des Végétaux Fossiles traversant les Couches du Terrain houiller, par M. Alex. Brongniart, à Paris, 1821.

17. VERTICAL STEMS IN COAL MINES.—Many instances of this phenomenon occur in England, several of which are noticed by Mr. Witham, a gentleman who has so materially contributed to our knowledge of the structure of fossil plants.* In the Derwent mines, at the depth of fifty-five fathoms, among numerous examples which were lying in horizontal layers, were several in an upright position. Two stems of sigillariæ were situated in the space cleared out to get at the lead ore, and stood erect, having their roots firmly impacted in a bed of bituminous shale ; they were about five feet high, and two in diameter. In the Newcastle coal-field, in a bed of sandstone 150 yards below the surface, are many erect stems of plants, having their roots in a thin bed of coal, as in this figure (Tab. 65). These plants are from two to eight feet in circumference.

18. TRUNKS OF CONIFEROUS TREES IN CRAIGLEITH QUARRY.—In the quarry of Craigleith, near Edinburgh, at a depth of 140 feet, an enormous trunk of a tree was discovered about ten years since. The length was thirty-six feet, and the circumference of the base nine. It lay in a nearly horizontal position, corresponding with the strata of sandstone in which it was imbedded. By polished transparent sections of the trunk, Mr. Witham was enabled to ascertain that this tree belonged to

* Observations on Fossil Vegetables, by Henry Witham, Esq. F.G.S. 1 vol. 4to. With Plates of the Internal Structure of Fossil Plants. Edinburgh, 1831.

the coniferæ. A few years afterwards, another tree
was found in this quarry. It was fifty-nine feet
long, and lay at an angle of about 40°, traversing
ten or twelve strata of sandstone. As is common
in these fossils, the trunk was crusted with coal,
probably the bark in a carbonized state.* In this
beautiful section of the stem, from Craigleith, presented by my lamented friend, the late Dr. Henry,
of Manchester, the coniferous structure is clearly
displayed ; and from this fact alone, the botanist
can delineate the general form and foliage of the
original tree, in like manner as the anatomist, from
a few fragments of teeth and bones, can determine
the affinities of the animal to which they belonged.

19. POLISHED SECTION AND MICROSCOPIC EXA-
MINATION OF FOSSIL TREES.—The discovery of a
process by which the structure of fossil vegetables
can be examined, with as much facility as that of
recent plants, has shed an unexpected light on the
ancient botany of our globe. On this plate of glass
you perceive a thin film of a dark colour, apparently of varnish. It is a slice of the darkest jet,
and if held between the eye and the light, is of a
rich brown colour, and displays a ligneous structure, resembling that of deal or fir ; it is, in fact,
a thin section of fossil coniferous wood ; for jet is
nothing more than the wood of some species of fir
or pine, that has undergone the process of bitumi-

---

* This specimen is represented by Mr. Fairholme in the
situation it occupied in the quarry.

nization, which I shall presently explain. When viewed under a microscope, the small glands, which I have mentioned as peculiar to the coniferæ (Tab. 63, fig. 2), are distinctly visible. The other specimens I now place before you are of silicified woods, prepared in the same manner. A few words, in explanation of the mode by which sections of such extreme thinness are obtained, may not be uninteresting. A slice is first cut from the fossil wood by the usual process of the lapidary. One surface is ground perfectly flat, and polished, and then cemented to a piece of plate-glass by means of Canada balsam. The slice thus firmly attached to the glass is now ground down to the requisite degree of tenuity, so as to permit its structure to be seen by the aid of the microscope. My specimens, as you perceive, are reduced to mere films or pellicles.* It is by this ingenious process that the intricate structure of any fossil plant can now be investigated, and the nature of the original determined, with as much accuracy as if it were now living.

20. NATURE OF COAL.—I now advance to the examination of the substance called *coal*, which is a mass of vegetable matter, transmuted by chemical changes into carbon, and still exhibiting the structure of the plants from which it was derived. When sections of coal, obtained by the process above described, are seen through the microscope, the fine,

* There are beautiful figures of these objects in Mr. Witham's work; and also in Dr. Buckland's Bridgewater Essay.

reticulated structure of the original is distinctly
visible, the cells of which are filled with a light,
amber-coloured matter, apparently of a bituminous
nature, and so volatile as to be readily expelled by
heat, before the texture of the coal is destroyed.*

Mr. Parkinson, whose work abounds in most
interesting observations and experiments on the
fossilization of vegetable substances, has shown
that the production of coal has depended upon a
change, which all vegetable matter undergoes when
exposed to heat and moisture, under circumstances
that exclude the air, and prevent the escape of the
more volatile principles.†   In this condition, a fermentation, which he terms the bituminous, takes
place, of which the phenomenon exhibited by *mow-
burnt hay* is a familiar example.   The production
of sugar, and, by continuance of the process, of
vinegar, is effected by vegetable fermentation in
the open air.   In the process of hay-making, the
saccharine fermentation is induced, and the grass
acquires a peculiar fragrance and sweetness; but
in wet seasons, when the hay is prematurely heaped
together, the volatile principles cannot escape from
the inner mass of vegetable matter, heat is rapidly
evolved, a dense vapour exhales, and at length
flames break forth, and the stack is consumed.

* Mr. Hutton.
† Organic Remains of a Former World, vol. i. p. 181. This
volume abounds in remarks and experiments on the fossilization of vegetables, of the highest interest.

When the process is interrupted, and combustion prevented, the hay is found to have acquired a dark-brown colour, a glazed or oily surface, and a bituminous odour. Were vegetable matter, under the circumstances here described, placed beneath great pressure, so as to confine the gaseous principles, bitumen, lignite, or coal, might be produced, according to the various modifications of the process. Mr. Parkinson traces vegetable matter through every stage of the saccharine, vinous, acetous, and bituminous fermentations; producing alcohol, ether, naphtha, petroleum, bitumen, amber, and even the diamond; and thus stems and branches have been converted into brown coal, lignite, jet, coal, and anthracite. A few of these bituminous substances require our attention.

21. MINERAL OIL, NAPHTHA, AND PETROLEUM. —I will first notice those bituminous fluids commonly known by the name of mineral oil. Springs or wells of this inflammable substance occur in many countries, as Persia, Calabria, Sicily, America, &c.; generally in rocks associated with coal. *Naphtha* is nearly colourless, and transparent, burns with a blue flame, emits a powerful odour, and leaves no residuum. Genoa is lighted with naphtha from a neighbouring spring. *Petroleum* is of a dark colour, and thicker than common tar; in some parts of Asia, this substance rises from coal-beds in immense quantities. From a careful analysis of petroleum, and certain turpentine oils, it is clear

that their principal component parts are identical;
and it appears therefore evident that petroleum has
originated from the coniferous trees, whose remains
have contributed so largely to the formation of coal;
and that the *mineral oil is nothing more than the*
turpentine oil of the pines of former ages—not only
the wood, but also large accumulations of the
needle-like leaves of the pines may also have contributed to this process. We thus have the satis-
faction of obtaining, after the lapse of thousands of
years, information as to the more intimate composition of those ancient destroyed forests of the period
of the great coal formation, whose comparison with
the present vegetation of our globe is the subject of
so much interest and investigation. The mineral
oil may be ranked with amber, succinite, and other
similar bodies which occur in the strata of the
earth. The occurrence of petroleum in springs
does not seem to depend on combustion, as has been
supposed, but is simply the result of subterranean
heat. According to the information we now possess, it is not necessary that strata should be at a
very great depth beneath the surface to acquire
a heat equal to the boiling point of water, or
mineral oil. In such a position the oil must have
suffered a slow distillation, and have found its way
to the surface; or have so impregnated a portion of
the earth, as to enable us to collect it from wells,
as in various parts of Persia and India.*   The

* Dr. Reichenbach.

author of an interesting paper in the American Journal of Science, remarks that petroleum is now daily discharging into the soft mud and gravel in the beds of the Muskingum and Hews's rivers. At Chilley, in this county (Sussex), beds of shanklin sand are permeated throughout with bituminous oil, originating either from neighbouring peat-bogs,* or from lignite beds of the Wealden.

22. BITUMEN, AMBER, AND MELLITE.—Bitumen may be described as an inspissated mineral oil; it is generally of a dark-brown colour, with a strong odour of tar.   In the Odin mine of Derbyshire, a species occurs which is elastic, being of the consistence of thick jelly, and bearing some resemblance to soft India-rubber; as it will remove the traces of a pencil, it has been named mineral caoutchouc. Some specimens possess the colour and transparency of amber: the soft bitumens may be rendered solid by heat.

From the bituminous substances which I have placed before you, to *Amber*, we pass by an easy transition ; for black amber bears, both in its appearance and composition, a close resemblance to the solid bitumens.   The nature of common amber is too well known to need remark; its electrical properties, odour, combustion, and the fact of its inclosing insects, leaves, and other foreign bodies, indicate its origin and former condition.   This substance is found in nodular masses, which are some-

* Fossils of the South Downs, p. 76.

times eighteen inches in circumference; it occurs in beds of lignite, and on the coast of Prussia in a subterranean forest, probably of the newer tertiary epoch. Mr. G. B. Sowerby mentions having seen, at Baden, the branch of a tree converted into jet, and having the centre filled with amber.* In the brown coal of Muskaw, amber occurs in the fossil coniferous wood, partly in disseminated portions, and *partly in the resin-vessels themselves;* and fir-cones are frequently discovered which contain this substance on and between the scales. Amber has also been found in coniferous plants associated with ferns, in coal that is referred to the upper secondary formations. In fine, there can be no doubt that amber is an indurated resin, derived from various coniferous trees, and which occurs in a like condition in all zones, because its usual original depositories, the beds of brown coal, have been formed almost everywhere under similar circumstances.†

A mineral substance, called *Mellite*, or honey-stone, from its colour, is found among the bitumi-nous wood of Thuringia.‡ In its chemical composi-tion, and electric properties, it bears a great analogy to amber; it is usually crystallized in small octahe-drons. In the tertiary beds of Highgate a fossil resin, resembling copal, has been discovered.

23. THE DIAMOND.—The chemical constituents

* Phillips's Mineralogy, p. 374.
† M. Goppert; Jameson's Edinburgh Journal.
‡ Organic Remains of a Former World, vol. i. Pl. 1, fig. 2.

of the substance I have described are chiefly carbon
or charcoal, and hydrogen, with a small proportion
of oxygen—the essential characters of vegetable
matter. In the diamond we have the elements of
pure carbon ; at a heat less than the melting point
of silver, it burns, and is volatilized, yielding the
same elementary products as charcoal. Sir Isaac
Newton long since remarked, that the refractive
power, that is, the property of bending the rays of
light, was three times greater in respect of these
densities, in amber and in the diamond, than in
other bodies ; and he therefore concluded that
the diamond was some unctuous substance that had
crystallized. Sir David Brewster has observed, that
the globules of air (or some fluid of low refractive
power) occasionally seen in diamonds, have communicated, by expansion, a polarizing structure to
the parts in immediate contact with the air-bubble,
a phenomenon which also occurs in amber. This is
displayed in four sectors of polarized light encircling
the globule of air ; a similar structure can be produced artificially, either in glass or gelatinous masses,
by a compressing force propagated circularly from a
point. This cannot have been the result of crystallization, but must have arisen from the expansion
exerted by the included air on the amber and the
diamond, when they were in so soft a state as to be
susceptible of compression from a very small force ;
hence Sir David Brewster concludes that, like
amber, the diamond has originated from the con-

solidation of vegetable matter, which has gradually
acquired a crystalline form by the slow action of
corpuscular forces.* The matrix of the diamonds of
Southern India is the sandstone breccia of the clay-
slate formation. Capt. Franklin observes that in
Bundel Kund, diamonds are imbedded in sandstone,
which he supposes to be the same as the new red
sandstone, for there are at least 400 feet of that
rock below the lowest diamond beds, and strong indications of coal underlying the whole mass.†

24. ANTHRACITE, CANNEL COAL, PLUMBAGO.
—The coal commonly used for domestic purposes in
this country is bituminous coal; containing, as
before stated, a volatile, inflammable fluid, in a cellular structure. The stone-coal, or anthracite,‡ as it is
termed, appears to be coal deprived of its bitumen;
for it is well known, that when basalt is in contact
with coal, the latter is in the state of anthracite; and
in some instances is even converted into plumbago,
the substance of which black-lead pencils are constructed. Anthracite generally occurs in rocks of an
earlier date than those which are strictly comprised
in the carboniferous group; but it is convenient to
notice the nature of the rock in this place, in connexion with the substance of whose vegetable nature
no doubt can exist. By a series of interesting experiments, Dr. MacCulloch has shown that there is a

* Geological Transactions, vol. iii. p. 459.
† London and Edinburgh Journal, October 1835.
‡ Anthracite, derived from the Greek, and signifying carbon.

natural transition from the bitumen to plumbago.
Hydrogen predominates in the fluid bitumen; bitumen and carbon in coal: in anthracite, bitumen
is altogether wanting; and in plumbago, the hydrogen also has disappeared, and carbon only, or
chiefly, remains.   With this general explanation of
the various states in which carbonized vegetable
substances occur, I pass to the consideration of the
process of petrifaction, that wonderful operation by
which the most delicate animal and vegetable structures are converted into solid rock.

    25. NATURE OF PETRIFACTION. — In many
instances, we find a mere substitution of mineral
matter for the original animal or vegetable substance.   Such are those casts of sandstone, indurated clay, and other consolidated materials, which
bear the forms and impressions of organic bodies,
but possess neither the internal structure, nor any
vestige of the constituent substances of the original.
Casts and impressions of shells, of the stems and
leaves of plants, and of fish-scales ; the flints, which
derive their form from echinites, &c., are familiar
examples of this process.

    In genuine petrifactions a transmutation of the
parts of an organized body into mineral matter takes
place.   Patrin, Brongniart, and other philosophers,
suppose that petrifaction has frequently been effected
suddenly, by the combination of gaseous fluids with
the constituent principles of organic structure.   It
appears, indeed, certain, that the conversion into

silex both of animal and vegetable substances,
must, in the majority of instances, have been almost
instantaneous, for the most delicate parts, those
which would undergo decomposition with great rapidity, are often preserved; such, for instance, as
the capsule of the eye, the membranes of the
stomach, the soft bodies of mollusca; and in plants,
the cellular and vascular tissue, and even the pollen.
The fact of the silicification of trees in loose sand,
and of the bodies of mollusca in their shells, as in
these fossil oysters from Brighton, while neither the
sand in the one instance, nor the shells in the other,
are impregnated with silex, cannot be explained by
the infiltration of a silicious fluid into cavities left
by the decomposition and removal of the animal
substance. A combination of gaseous fluids, with
the constituent principles of the animal or vegetable, changing the latter into stone, without modify-
ing the arrangement of their molecules, so as to
alter their forms, seems the only mode by which
such a transmutation can have been effected.   The
production of congelation, by a simple abstraction
of caloric, is akin to this change; but petrifaction
is induced by the introduction of another principle.*
As to density, the most subtle gaseous fluids may
acquire the greatest solidity; as, for example, in the
union of oxygen with metallic substances. Oxygen
is supposed by Patrin to be a chief agent in the
phenomenon of petrifaction, by its combination

* See Pidgeon's Cuvier, on Fossil Animals.

with the phosphoric principle, which is present in organized bodies.

26. ARTIFICIAL PETRIFACTIONS.—Last year M. Goppert published the result of an interesting investigation of the condition of fossil plants, and the process of petrifaction. Mr. Parkinson had remarked, that the leaf in ironstone nodules might sometimes be separated in the form of a carbonaceous film; and M. Goppert having lately found similar examples, was induced to undertake a set of experiments. He placed fern leaves in clay, dried them in the shade, exposed them to a red heat, and obtained striking resemblances to fossil plants. According to the degree of heat, the plant was found either brown, shining black, or entirely lost, the impression only remaining; but in the latter case the surrounding clay was stained black, thus indicating that the colour of the coal shales is from the carbon derived from the plants they include. Plants soaked in a solution of sulphate of iron, were dried and heated till every trace of organic matter had disappeared, and the oxide was found to present the form of the plant. In a slice of pine-tree the punctured vessels peculiar to this family of vegetables were perceptible. These results by heat are probably produced naturally, by the action of moisture under great pressure, and the influence of a high temperature.

27. DIFFERENT STATES OF THE FOSSILIZATION OF WOOD.—A most valuable communication on "Wood partly petrified by Carbonate of Lime," has

recently been made to the Geological Society, by my
friend Charles Stokes, Esq.*   The specimen which
gave rise to these remarks was a piece of beech-
wood, from a Roman aqueduct in Germany, in which
were several insulated portions, converted into carbonate of lime, while the remainder was unchanged.
I cannot enter upon Mr. Stokes's interesting observations on this specimen, but his statement of the
different conditions in which fossil wood appears, is
highly important.   Sometimes the most minute
structure is preserved, as in the vessels of palms
and coniferæ, which are as distinct in the fossil
as in the recent trees.   From this state of perfection, we have every degree of change, to the last
stage of decay : the condition of the wood, therefore,
had no influence on the process.   The hardest wood,
and the most tender and succulent, as for instance,
the young leaves of the palm, are alike silicified.
In some instances, the cellular tissue has been petrified, and the vessels have disappeared ; here silicification must have taken place soon after the wood
was exposed to the action of moisture, because the
cellular structure would soon decay ; the process
was then suspended, and the vessels decomposed.
In other examples, the vessels alone remain ; a proof
that petrifaction did not commence till the cellular
tissue was destroyed.   The specimens where both
cells and vessels are silicified, show that the process
began at an early period, and continued till the

* Transactions of the Geological Society, vol. v. p. 207.

whole vegetable structure was transmuted into stone.* My lamented friend, Dr. Turner, in some admirable comments on the subject of petrifaction, remarks, that whenever the decomposition of an organic body has begun, the elements into which it is resolved are in a condition peculiarly favourable to their entering into new combinations; and that if water, charged with animal matter, come in contact with bodies in this state, a mutual action takes place, new combinations result, and solid particles are precipitated, so as to occupy the place left vacant by the decomposed organic matter.

Mr. Parkinson, in corroboration of his opinion that wood undergoes bituminization before it becomes petrified, mentions, that a specimen of wood from Walton, which was changed into marble, and took a beautiful polish, left, upon removing the carbonate of lime by muriatic acid, a mass of light, inflammable, bituminous wood, which possessed a volume almost equal to its original state.†

28. HAZEL-NUTS FILLED WITH SPAR.—Before I quit this subject I would notice a singular fact — an instance of partial mineralization in which mineral matter has permeated the shells of hazel nuts, without altering their structure, although the interior is lined with spar. In Belfast Lough, a bed of submarine peat is situated beneath the ordinary level of the waters, but is generally left bare at the

* Geological Transactions, vol. v. page 212.
† Organic Remains, vol. iii. p. 440.

ebb tides. Trunks and branches of trees are imbedded in the peat, and vast quantities of hazel-nuts,
the whole being covered by layers of sand, shells,
and blue clay. On the east side of the Lough, limestone rocks exist, and the nuts in the peat contain
calcareous spar. Some specimens are full, others
are only lined with groups of crystals. The shells
of these fruits are entire, and have undergone
no change, their substance being in the state of
common dried hazel-nuts. On the west side of the
Lough, the rocks are schistose, and the nuts, as is
common in peat, are empty. The specimens in my
collection, collected by Dr. Bryce, of Belfast, exhibit the different conditions I have described.

29. SILICIFICATION, or PETRIFACTION BY
FLINT.—The various forms in which silex is found,
are proved to have been dependent on its state of
solution; in quartz crystals it was entirely dissolved,
in agate and chalcedony it was in a gelatinous state,
assuming a spheroidal, or orbicular disposition,
according to the motion given to its molecules. Its
condition was also modified by the influence of
organic matter. In some polished slices of flints
from Bognor, presented by Sir Richard Hunter,
and from Worthing, by Frederic Dixon, Esq.* the
transition from flint to agate, chalcedony, and
crystallized quartz, is beautifully exhibited. The

* In Mr. Dixon's choice collection of chalk fossils, at Worthing, there is an interesting specimen of *Spherulites Mortoni*,
which will be figured in Mr. Lyell's Elements of Geology.

shell of an echinus, in my possession, is transmuted
into crystallized carbonate of lime, and the lower
portion of the cavity occupied by flint, the upper surface of the latter being covered by crystals of calcareous spar. The curious fact that the shells of the
echinites in the chalk are almost invariably filled
with flint, while the crustaceous covering is converted into calcareous spar, is, perhaps, attributable
to the animal matter of the echinus having undergone silicification ; for the most organized parts are
those which appear to have been most susceptible of
silicious petrifaction. In another specimen, in my
museum, the body of an oyster is turned into flint,
while the shell is, as usual, carbonate of lime.* The
shells of mollusca, the crustaceous skeletons of
echini, and the bones of the belemno-sepiæ, appear
to have possessed too little animal matter, and to
have been too much protected by calcareous earth,
to have become silicified ; they are changed into spar
by water charged with carbonic acid gas, having
insensibly effected the crystallization of their molecules.†

* See an interesting essay on this subject, by M. Alexandre
Brongniart, "*Essai sur les Orbicules Siliceux, &c.*" Paris, 1831.

† Mr. Reade's highly important experiments and observations on the structure of plants, appear to throw some light on
this interesting subject—the silicification of wood. " By the
agency of heat the surrounding silicious matter may be liquefied,
and the carbon and gaseous products of the wood dispelled,
while the essential characters of the fibrous and cellular structure are undisturbed. The unconsumed portions, which alone
constitute the true vegetable frame-work, are thus, as it were,

I have dwelt at considerable length on the processes by which animal and vegetable structures
have been mineralized, and preserved for countless
ages; but you will not, I trust, consider that this

mounted in the fluid silica. This property of vegetable fibre
of retaining its form, notwithstanding the action of a high
temperature, suggested to me the probability of detecting structure in the ashes of coal; and upon examination, I found that
the white ashes of ' slaty coal,' furnish most beautiful examples
of vegetable remains." In a subsequent paper the author adds
the following remarks :—" Having ascertained that the silicious
organization of recent plants is not destructible, even under the
blow-pipe, it appeared to me a natural inference, that the less
intense heat of a common fire would not destroy this silicious
tissue in the coal-plants; and my opinion has been confirmed, for
I have detected in the white ashes of coal all the usual forms of
vegetable structure, viz. cellular tissue, smooth and spiral fibre,
and annular ducts. A comparison of the ashes of coal with
those of recent plants, would doubtless afford some further in-
sight into the nature of fossil vegetables. To mention only one
instance—I have ascertained that the lumps of carbonized
matter, which occur abundantly in the upper sandstone near
the Spa at Scarborough, are, in all probability, portions of the
stems of some arundinaceous or gramineous plants. The structure of the epidermis is precisely similar to that of the oat,
consisting of parallel columns, set with fine teeth, dove-tailing,
as it were, into each other, while the underlying tissue consists
of cubical cells, a thin horizontal section exhibiting a series of
squares. From these facts it is evident, that the true frame-
work and basis of vegetable structure in the plants of coal, is not
only entirely independent of carbon, but that it has also resisted
the bituminous decomposition, which has converted all the carbonaceous materials into a highly inflammable substance."—*Rev.
J. B. Reade on the Structure of the Solid Materials found in the
Ashes of Recent and Fossil Plants.* Journal of Science, vol. ii.
p. 413.

deeply interesting inquiry has occupied too much of
your attention.   In an earlier stage of our geological
argument, I was unwilling to enter upon its consideration, lest our minds should not have been prepared
to comprehend, or relish an investigation of so
scientific a character.   I now hasten to call your
attention to the flora of the ancient world, entombed
in the carboniferous strata.

30. PLANTS OF THE COAL STRATA.—The beds
of coal, as we have already seen, are actually composed of fossil vegetable stems, branches, and leaves ;
and possibly the different kinds of coal may, as Mr.
Reade has suggested, have resulted from original
differences in the vegetables whence they were derived.   The coal shales, or slates, are highly charged
with carbon, a character which M. Goppert's experiments tend to elucidate ; and they contain, as I
have previously remarked, a profusion of fossil
plants.   The vegetable remains in the coal strata are,
I believe, universally in a carbonized state ; and the
leaves sometimes possess such a degree of tenacity
and elasticity as to be separable from the stone, as
Mr. Parkinson long since observed.   The leaves
and seed-vessels which occur in the iron-stone
nodules have, in many instances, undergone a metallic impregnation, as is exemplified in this splendid
series of specimens from Coalbrook Dale, for which
I am indebted to John Pritchard, Esq., of Brosely.
Brilliant sulphuret of iron, or pyrites, in some examples, permeates the entire vegetable structure ;

in others, the stems and leaflets are replaced by
white hydrate, or sulphate of alumina; and in many
by crystals of galena, or sulphuret of lead, and of
blende.* In the sandstones, the vegetable stems
have generally a carbonaceous crust, and their
structure is sometimes found in a calcareous or silicious state.†

The coal plants, which have been accurately determined, amount to upwards of three hundred; of
these, two-thirds are related to the arborescent ferns,
and the higher tribes of the cryptogamia; about
ten species to the flowering monocotyledonous trees;
as many to the coniferæ, and cactaceæ; and numerous species still remain undescribed.   I will now
place before you a few of the usual forms that
occur in the British coal-measures.

31. Fossil Mare's-tail, or Equisetum.—The
mare's-tail (*equisetum fluviatile*) of our marshes and
ditches, is an elegant plant, having a succulent,
erect, jointed stem, with attenuated foliage growing
in whorls around the joints, the latter being protected by a distinct striated sheath; the parts of
fructification constitute a scaly catkin at the apex
of the stem.   There are ten species of this genus,
eight of which are natives of England; the stem
of the largest does not exceed half an inch in diameter.   In the coal-measures remains of this genus

---

* Sulphuret of zinc—*blende* is a German term, signifying
glistening.

† *Vide* Organic Remains of a Former World, vol i.

occur in abundance, and are referrible to many gigantic species; some of the stems are fourteen inches in diameter; the *equisetum columnare*, a fossil of constant occurrence in the carboniferous strata, is beautifully figured and illustrated by M. Ad. Brongniart.*

32. FOSSIL FERNS (Tab. 64).—The *brake*, or fern, of our commons and waste lands, is a familiar example of a remarkable and numerous family of plants, distinguished by the peculiar distribution of the seed-vessels. The arborescent ferns rise into trees from thirty to forty feet in height, their stems being marked with scars from the decay of the leaf-stalks, and their summits covered with an elegant canopy of foliage; their general appearance is shown in this sketch (Tab. 69, fig. 5). The leaves of the smaller species are very elegant, and present immense variety in their forms and in the modes of distribution of the veins of the leaf; from the character of the latter, M. Adolphe Brongniart has formed the generic distinctions of the fossil plants of this family. The beautiful state in which these remains occur, is shown in the numerous specimens before us (see Tab. 64, figs. 2, 3, 4). The fructification on the back of the leaf is sometimes distinctly visible.

The stems, with their elliptical cicatrices, or scars, bear some resemblance to those of the palms; but are readily distinguished from their longest diameter

* Végétaux Fossiles, tom. i. pl. 13.

1   2

3   4

TAB. 64.—PLANTS OF THE COAL-MEASURES.

Fig. 1. *Asterophyllites Parkinsoni, in coal strata.* 2. *Pecopteris Mantellii.* 3. *Sphenopteris linearis.* 4. *S. affinis, from Burdie House, by Dr. Hibbert.*

being vertical, while in the palms it is transverse : sections of the stems of these two tribes have also distinctive characters.* The large tree-ferns are confined almost exclusively within the tropics; humidity and heat being the conditions most favourable to their development. In the coal, there are not less than 130 known species of ferns, nearly all of which belong to the tribe of polypodiaceæ ; the common polypody, so frequent on the walls of old buildings, will convey an idea of the general character of the foliage. In speaking of the stems of ferns, I must remind you of the fossil plant from the Wealden, the *clathraria Lyellii* (page 342), in which the scorings on the outer surface, from the removal of the petioles, bear an analogy to those of the stems of tree-ferns and palms ; but the internal axis, so well shown in the specimen (Tab. 44), separates it from those families.

33. SIGILLARIA † (Tab. 65).—Among the most common fossils that strike the attention of the observer in coal mines, are long, flat, fluted slabs, marked with impressions, disposed with great regularity, and presenting an infinite variety of patterns. These are the compressed, cylindrical stems of plants related to the arborescent ferns ; the imprints being the scars left by the leaf-stalks. No foliage has been found attached, but it is probable that many of the leaves so abundant in the shales belong to

* Végétaux Fossiles, tom. i. pl. 37.
† So named from the impressions on the surface.

them.   The stems vary from a few inches to three
feet in circumference, and specimens have been discovered that indicate a length of sixty feet.

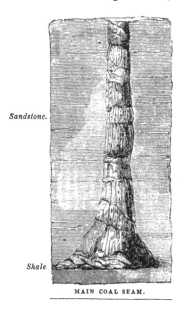

Sandstone.

Shale.

MAIN COAL SEAM.

TAB. 65.—SIGILLARIA PACHYDERMA.

(*Drawn by Miss Ellen Duppa.*)

The stems often escape compression, stand perpendicularly, intersecting the horizontal strata, and
sometimes have roots proceeding from the base.
They are generally surrounded by an envelope, an

inch in thickness, of fine, crystalline, bituminous coal. The longitudinal platings, which are the characteristic marks of the sigillariæ, are commonly indistinct at the base. A specimen figured in the beautiful and highly interesting work of Dr. Lindley and Mr. Hutton,* was ten feet high (Tab. 65), and two feet in diameter at the base. Its roots were in shale, immediately above a bed of coal, and the trunk extended through several strata of shale and sandstone. The sigillariæ were evidently hollow, like the reed, and with but little substance, as is proved by the extreme thinness of the specimens, which lie in a horizontal position, and are compressed. The upright stems consist entirely of sandstone, within the envelope of coal. Nearly fifty species are enumerated by M. Brongniart. The original trees appear to have been closely related to the arborescent ferns, but their leaves were small, and differently disposed than in any existing species.

Stems of several other genera of plants are found in the coal, the form and distribution of the markings of the leaves being very dissimilar to those I have described.†

34. LEPIDODENDRON (Tab. 66). — The most elegant and abundant of the fossil plants of the coal are the lepidodendra, so named from the scaly appearance of the stems, produced by the separation of the leaf-stalks. The scars are simple, lanceolate,

* The Fossil Flora of Great Britain.
† See Dr. Buckland's Bridgewater Essay, vol. ii.

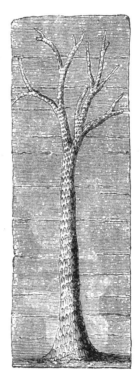

TAB. 66.—THE LEPIDODENDRON STERNBERGII.
Thirteen and a half feet wide, and thirty-nine feet high.
(*Drawn by Miss Ellen Duppa.*)

This figure, from the Fossil Flora of Great Britain,* represents a specimen
nearly forty feet in length, which was discovered in the Bensham coal
seam, in the Jarrow coal-field. The width of the base was thirteen and
a half feet.

* Plate 203.

rhomboidal, and arranged spirally round the stem ;
the latter is slight and tapering, and sometimes
arborescent. The cones, or stroboli, so common in
the ironstone nodules, are the fruit of these elegant
trees. In the markings on the stems, the Lepidodendra resemble the club-mosses (*Lycopodiaceæ*),
which are herbaceous, prostrate plants, found in
damp woods and bogs, having their leaves simple
and imbricated—that is, lying over each other ; the
tropical species, which are the largest, do not
attain a greater height than three or four feet. In
their structure and seed-vessels, the lepidodendra
approach the coniferæ.

Count Sternberg remarks, that we are unacquainted with any existing species of plant, which,
like the lepidodendron, preserves at all ages, and
throughout the whole extent of the trunk, the scars
formed by the attachment of the petioles, or leaf-
stalks, or the markings of the adhesion of the leaves
themselves. The yucca, dracæna, and palm, entirely
shed their scales when they are dried up, and there
only remain circles, or rings, arranged round the
trunk in different directions. The flabelliform palms
preserve their scales at the inferior extremity of
the trunk only, but lose them as they increase in
age ; and the stem is entirely bare, from the middle
to the superior extremity. But in the lepidodendron, the scales follow a decreasing proportion from
the base of the trunk to the extreme branches.*

* " Faut-il ranger ce genre d'arbres parmi un groupe de

Dr. Lindley and Mr. Hutton describe these plants as constituting a gradation from the flowering to the flowerless tribes.

35. Fossil Club-Moss (Tab. 67).—In a specimen from the Tyrol, with the precise locality of

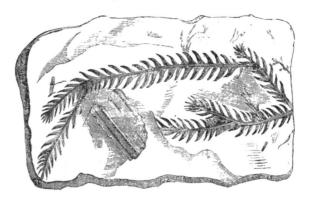

TAB. 67.—LYCOPODITES BENETTIÆ.*
(*Fossil Club-Moss.*)

(*Drawn from nature, by Miss Ellen Duppa.*)

which I am unacquainted, the characters of the lycopodiaceæ are beautifully displayed. The plant

végétaux, qui aujourd'hui est indigène seulement entre les tropiques, et dont la végétation exige un climat plus doux et une température plus élevée?"—*Sternberg*, Flora zur Vorwelt.

* This elegant plant is named in honour of Miss Etheldred Benett, of Norton House, Wilts ; a lady, whose liberal contributions of specimens, and instructive observations on the chalk fossils of Wiltshire, afforded me important assistance in my early attempts to investigate the organic remains of Sussex.

is in the state of carbon, or rather of indurated bitumen, and its dark tint admirably contrasts with the cream-coloured shale in which it is imbedded. Several species of lycopodites are described as occurring in the coal of Newcastle, and of Silesia.

36. STIGMARIA.—Very few traces of any but land plants have been observed in the coal strata; there is, however, one remarkable exception. A frequent fossil in the Derbyshire strata is a compressed subcylindrical stem, having the surface spirally studded with tubercles, and containing internally an imbricated body or core. The late Mr. Martin* figured some illustrative examples of this plant under the name of *Phytolithus imbricatus:* but their true characters were unknown, till Dr. Lindley and Mr. Hutton discovered specimens, in which the stems were united to a trunk. From the observations of these gentlemen, it appears that the plant had a central trunk, of a compressed sub-conical form, the substance of which, from the corrugated surface in the fossils, is supposed to have been pulpy or soft. From this central body proceeded from ten to fifteen branches, disposed horizontally, and dividing at unequal distances; when perfect, it is computed they would have extended twenty or thirty feet. The fossil stems are fragments of these branches, the tubercles being the bases of cylindrical, succulent leaves, several feet in length; the internal axis of which I have

* Petrificata Derbiensia.

P P

spoken, like that of the *Clathraria Lyellii* (page 341), being the woody pith or core. The external hollow cylinder is entirely composed of spiral vessels. The stigmaria was an aquatic plant, inhabiting swamps or lakes, and resembling the euphorbiaceæ in its internal structure.*

37. SEED-VESSELS.—Seed-vessels occur in the carboniferous strata, and one species is not un-

TAB. 68.—SEED-VESSELS FROM COAL.

(*By W. D. Saull, Esq. F.G.S.*)

common, but its natural affinities have not been determined. Among the numerous plants which I have omitted to notice, in this rapid sketch of the flora of the coal, is an extinct genus that differs altogether from existing types. It is called *asterophyllites*, from the verticillate arrangement of the foliage (Tab. 64, fig. 1), and is supposed to belong to the dicotyledonous class.

38. CONIFERÆ.—Our previous investigation of the structure of the recent coniferæ (page 544) renders it unnecessary to explain the manner in

* Figures and descriptions of this plant are given by Dr. Buckland, Plate 56, page 477.

which the nature of the fossil stems and woods can be determined. Mr. Witham, by microscopic observations on polished sections of fossil woods, has proved the existence of several species of coniferous trees in the coal strata of Scotland; and similar examples have since been discovered in other carboniferous deposites. It is not a little curious that all the species are related to the Auracaria, or Norfolk Island pine, and not to the common coniferæ; this is proved by the ducts of the vessels being arranged alternately, and in double and triple rows, (see page 543). The pines of the coal have but few and slight appearances of the lines by which the annual layers are separated, and resemble in this respect the existing species of tropical regions; we may therefore infer that the seasons of the countries where the coal-plants flourished were subject to as little diversity, and that the changes of temperature were not abrupt.* It is said that in the coal of Nova Scotia and New Holland, coniferæ with the ordinary structure occur.

39. REVIEW OF THE CARBONIFEROUS FLORA.— My limits forbid a more extended notice of the fossil plants of the carboniferous system, and I will now take a summary review of the principal facts that have been submitted to our notice. We have seen that the striking character of the flora of that incalculably remote epoch, is the immense numerical ascendance of the vascular, or higher tribes of

* Witham.

cryptogamic plants, which amount to two-thirds of
the whole of the species hitherto determined.
With these were associated a few palms, coniferæ
allied to the auracariæ, and dicotyledonous plants
approaching to the cacteæ, and euphorbiaceæ.
The vast preponderance and magnitude of the vegetables bearing an analogy to the tribes of *ductulosæ*,
but differing from existing species and genera,
constitute the most remarkable botanical feature :
thus we have plants related to the mare's-tail, (*calamites*) eighteen inches in circumference, and ten or
twelve feet high : tree-ferns (*sigillariæ*) fifty feet
in height ; and arborescent club-mosses (*lepidodendra*) attaining an altitude of sixty or seventy
feet. The contrast which such a flora presents to
that afforded by the woods and forests of dicotyledonous trees, and the verdant turf, which now grow on
the surface of the carboniferous districts of England,
is as striking as the discrepancy between the zoology
of the secondary formations and that of the present
day. The restoration of some of the vegetable forms
which flourished in the carboniferous era, will perhaps prove more illustrative of this phenemenon
than mere description. (See Tab. 69.)

To arrive at any satisfactory conclusions as to the
nature of the countries which supported the plants
of the carboniferous strata, we must consider the
geographical distribution of the related existing
genera, and the circumstances which conduce to their
full development. It is now well known that a hot

climate, humid atmosphere, and the unvarying temperature of the sea, are the circumstances which

1.   2. 3. 4.   5. 6. 7. 8.   9.

TAB. 69.—THE FLORA OF THE CARBONIFEROUS EPOCH.

(Designed and drawn by Miss Ellen Maria Mantell.)

Fig. 1. *Auracaria.* 2. *Asterophyllites comosa.* 3. *Pandanus.* 4. *Equisetum.* 5. *Arborescent fern.* 6. *Fern.* 7. *Calamites.* 8. *Lepidodendron.*
9. *Sphenopteris.*

exert the most favourable influence on the growth
of the ferns and other cryptogamic plants ; low
islands in tropical latitudes being the localities

where these forms of vegetation flourish most luxu-
riantly.   We may therefore infer that the plants
entombed in the coal strata, grew in a climate of at
least as high a temperature as that of the tropics;
and probably in insular situations; thus we obtain
evidence of the existence during the carboniferous
epoch, of a tropical Polynesia clothed with forests
of palms, tree-ferns, and gigantic equisetaceæ and
lycopodiaceæ.

40. FORMATION OF NEW COAL-MEASURES.—
Let us now inquire what were the circumstances
which gave rise to these prodigious layers of carbonized matter, unmixed with other materials—
these immense beds of vegetables, from which animal
remains appear to have been almost wholly excluded—and whether accumulations of vegetables,
which in after ages shall present phenomena of a
like nature, are now in progress?   We have seen
that the plants in the coal-measures are for the most
part lying horizontally, as if whole groves and forests
had been laid prostrate, and become matted together, the small and more delicate tribes being
entangled in the general mass; presenting, in fact,
a very similar arrangement to that observable in the
subterranean forests, peat-bogs, and other modern
accumulations of vegetables.   If we extend our
view to operations which are now going on in
countries covered with a dense vegetation, abounding in lakes and marshes, and traversed by vast
rivers, we shall no longer feel surprised at the

immense quantity of vegetable matter of which the
coal-measures are composed.   In America, trees in
prodigious quantities, the wrecks of whole forests,
are borne down by the tributary streams into the
great rivers, and hurried along by the mighty flood
of waters, till arrested in their course they become
entangled, and form stationary masses called rafts,
which in the Mississippi, and other large rivers of
North America, extend over many leagues, and are
of great depth : in some instances particular species
are associated together, as cedars, pines, and firs,
without the intermixture of other trees.   Near the
mouth of the Mississippi, immense rafts, composed
of drifted trees brought down every spring, constitute a matted bed of vegetables, which is many
yards in thickness, and stretches over hundreds of
square leagues.   These rafts become covered with
fine mud and sand, on which other trees and plants
are drifted down the following year ; earthy deposites again take place, and thus alternations of
vegetables with layers of calcareous matter are
annually produced.*

In the lower plain of the Mississippi, immense
inundations continually occur from the melting of
the snow, and the flood of water thus suddenly
poured into the bed of the river, and that of the
Missouri. " The mouths of the large tributary rivers
are thus absolutely choked up, and their waters,
being driven backwards, overflow their banks, and

* *Vide* Principles of Geology, vol. i.

inundate the lower parts of the plain, forming lakes
of twenty miles or more in length. Here we have the
conditions required for the formation of future coal
fields : rapid development of vegetation in swamps,
and periodical inundations of water charged with
mud and sand."* That the bituminization of
vegetables, and their subsequent consolidation by
pressure into coal, might take place under the conditions here contemplated, we can readily conceive ;
but in the coal-measures we have vast alternations
of strata that abound in marine remains. But rafts
might be drifted into the tranquil depths of the
ocean, and become covered with mud and sand;
and a repetition of this process, at intervals, during
a long period of time, would be sufficient to produce
the appearance described. The occasional vertical
position of the stems of trees, and the admirable
preservation of delicate leaves, do not appear to me
to invalidate this inference; for in the entangled
masses in the American rivers, trunks of trees often
occur upright ; and my distinguished friend, Sir
Edward Codrington, informed me, that in the interior of the rafts, grasses and tender plants are
found entire. Such masses, therefore, might be
drifted thousands of miles, and yet the imbedded
fragile species, protected by the external network of
entangled branches, remain uninjured; and, undergoing bituminization, enveloped by the soft mud
permeating the mass, become changed into those

* Bakewell's Geology.

durable forms which abound in the natural herbaria of the carboniferous strata.

41. CORALS AND CRINOIDEA OF THE CARBONI-FEROUS SYSTEM.—But I must pass to the consideration of the animal remains of the strata, which have afforded so rich and varied a field of botanical research. From the examination of the fossil corals and crinoidea in the previous lecture, a few remarks on this subject will suffice. About thirty species of polyparia occur in the limestones, but no traces have been observed in the coal-measures. They consist principally of species of tubiporæ and cateniporæ (page 502, Tab. 58); cyathophylla (page 561, Tab. 57, figs. 4, 8); astreæ (Tab. 57, fig. 10); turbinoliæ (page 57, figs. 1, 3); and fungiæ (Tab. 57, fig. 2). Of the crinoidea, more than thirty species have been found. The mountain-limestone is the grand depository of the encrinites, and entire beds of marble are formed of their fossilized skeletons, as I explained in the former discourse (page 512). One singular genus of the crinoidea, *pentremites*, occurs in the carboniferous limestone (Tab. 53, fig. 8), of which there are several specimens in my museum collected by the late Lady Crewe, of Calke Abbey; and I have also species of the same genus from the United States, by Professor Silliman. One species of cidaris (*C. Phillipsii*), with large mammillated tubercles, and muricated spines, has been found, and is the earliest geological appearance of the family of echinodermata.

42. Shells of the Carboniferous System. —The testaceous remains of above two hundred species of the various families of mollusca have been' determined. The ancient forms of terebratulæ (*brachiopodous mollusca*), which first appeared in the saliferous strata (page 421) abound in the mountain limestone; and in some localities the rocks are formed of spiriferæ, productæ, and terebratulæ, conglomerated by calcareous cement. Bivalves comprising about ten fluviatile, or fresh-water species, occur in some of the coal-measures, forming regular layers, called by the miners *muscle-bands*, from the character of the shells (uniones), of which they are chiefly composed. The marine tribes are in a great measure confined to the limestone below the coal. But in Yorkshire, Professor Phillips discovered a remarkable exception ; in the coal-measures of that county, there is a thin layer of marine shells, intercalated between strata of fresh-water. The nautili, ammonites, and other cephalopoda, amount to sixty species ; and two very remarkable genera of this order appear : — the bellerophon (Tab. 53, fig. 5), and the orthoceras, or orthoceratite (Tab. 53, fig. 12). The latter may be described as a straight nautilus ; it is a conical, chambered shell, having entire septa, pierced with a siphunculus ; a reference to the remarks on the fossil nautilus will explain the nature and functions of this structure (see page 291). These shells are often of considerable magnitude ; a specimen from Sweden, in my collection, is above

twenty inches in circumference ; and others are described as being two feet and a half in length, with
sixty-six septa.   Marbles, formed of these remains,
also occur ; and polished slabs afford interesting sections of the septa and siphunculus.

43. CRUSTACEA.—In reviewing the zoological
characters of the strata, in an ascending series, the
remains of crustaceous animals next arrest our
attention ; and so many extraordinary forms of
this family are met with in the carboniferous rocks,
that in order to exemplify their nature, I shall offer
a few remarks on the structure and economy of the
existing species.

Crabs and lobsters are familar examples of this
class of animals, whose skeletons are external, and
whose circulation, respiration, and organs of loco-
motion, are very peculiar.   They occur in a fossil
state in the tertiary formations, as I have already
mentioned ; extinct species of lobsters, crabs, &c.
being found in the beds of London clay near
the metropolis, and in the isle of Sheppey.   In the
chalk, crustacean remains are comparatively rare ;
but my collection contains some beautiful examples of
astacidæ allied to the cray-fish, from the South Downs ;
and several species and genera from the Galt.*

The Wealden exhibits no traces of this family,
with the exception of the minute cases or shields of
the cypris (page 344), which so largely contributed to the formation of the Sussex marble.   In

* Geology of the South-East of England, p. 169.

the beds of the oolite and lias, in some localities, their remains are profusely scattered; the lithographic slate of Solenhofen alone have afforded to the researches of Count Munster nearly fifty species (page 399).

The crustacea, like all other tribes which are destined to live in water, perform respiration by certain external organs, termed *branchiæ*, formed by a peculiar modification of the external covering; these organs present great variety of form and

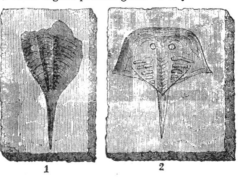

1		2

TAB. 70.—CRUSTACEA OF THE CARBONIFEROUS LIMESTONE.
Fig. 1. *Asaphus caudatus.* 2. *Limulus, from Coalbrook Dale,*

disposition, according to the habits and economy of the different species. In some kinds, as the crab and lobster, the branchiæ are fixed to the sides of the thorax, and inclosed in especial respiratory cavities; these organs consist of many thousand minute filaments, like the fibres of a feather; and to these are attached short and flat paddles, which are kept in incessant motion by proper muscles, and thus the water is agitated, and its full action on the branchiæ maintained.

44. The Limulus.—In another division, the *Limulus*, or king-crab—of which a beautiful specimen, presented by H.Y. Everitt, Esq., is before you—

TAB. 71.—TRILOBITE FROM DUDLEY.

(*Calymene* * *Blumenbachii.*)

a genus abundant in the seas of India and America, the gills are disposed on lamelliform processes. The limulus has a distinct carapace or buckler, with two eyes in front of the shield. A small fossil

* Calymene, *concealed;* alluding to the non-discovery of legs or antennæ.

species (Tab. 70, fig. 2) is found in the iron-stone nodules of Coalbrook Dale.

45. TRILOBITES.*—In the carboniferous strata we find those extraordinary forms of animal organization, which, under the names of Dudley locusts and trilobites, have long been known to naturalists and collectors. They belong to a family of the crustacea, which appears to have become extinct after

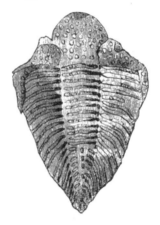

TAB. 72.—TUBERCULATED TRILOBITE, FROM DUDLEY.

(*Asaphus† tuberculatus.*)

the deposition of the coal, no traces of their remains having been discovered in rocks of a more recent period. The trilobites have been divided into ten

* *Three-lobed*, from their general form.
† Asaphus, *obscure.*

or twelve genera, comprising nearly sixty species, and to which additions are constantly being made. Their general structure consists of a crustaceous shell, divided by longitudinal grooves, or furrows, into three lobes, (hence the name,) and having segments which were capable of being folded over each other. Some species (Tab. 71) could coil themselves into a ball, like the millepedes; others had the central segments alone moveable (Tab. 72); and many were furnished with a tail, or post-abdomen (Tab. 70, fig. 1). No traces of legs have been discovered, and it is, therefore, inferred that they possessed soft, perishable paddles, bearing branchiæ. The only recent genus, that possesses any affinity to these animals, is the *Serolis*, several species of which inhabit the Straits of Magellan, and are described by Dr. Buckland; but these have legs and antennæ fully developed.

46. EYES OF THE TRILOBITE.—The structure of the eyes of the crustacea is similar to that of insects, the nature of which was fully explained in my lectures on the nervous system. It will be sufficient to state, that the eyes of these creatures are immoveable, and that this apparent deficiency is compensated by a visual organ of a most extraordinary kind. The eye is made up of a vast number of elongated cones, each having a crystalline lens, pupil, and cornea, and terminating on the extremity of the optic nerve.* Each organ of sight

* See Dr. Roget's illustration of this subject, vol. ii.

is, therefore, a compound instrument, made up of a series of optical tubes, or telescopes, the number of which in some insects is quite marvellous. Thus each eye of the common house-fly is composed of eight thousand distinct visual tubes ; that of the dragon-fly, of nearly thirteen thousand ; and of a butterfly, seventeen thousand. That any traces should remain of the visual organs of a crustaceous animal, which must have perished at so remote a period, seems at first incredible ; but there are no limits to the wonders which geology unfolds to us.* The trilobite, like the limulus, was furnished with two compound eyes, each being the frustrum of a cone, but incomplete on that side which is opposite to the other eye. In the asaphus (Tab. 72), four hundred spherical lenses have been detected in each eye; but in general the lenses have fallen out, as often happens after death in the eyes of the common lobster. "Thus," observes Dr. Buckland, "we find in the trilobites of these early rocks, the same modifications of the organ of sight as in the living crustacea. The same kind of

* This structure of the eye of the trilobite was, I believe, first noticed by that accurate observer, Mr. Martin, the author of Petrif. Derbiensia. In the work of my friend, M. Brongniart (Histoire Naturelle des Crustaces Fossiles, par A. Brongniart et G. A. Desmarest, 1 vol. 4to, with Eleven Plates, Paris, 1822), the eye of the trilobite is beautifully represented. In Dr. Buckland's Bridgewater Essay, this subject, like every other which Dr. B. investigates, is ably elucidated, and placed before the reader in a striking point of view.

instrument was also employed in the intermediate
periods of our geological history, when the secondary strata were deposited at the bottom of a sea
inhabited by limuli, in the regions of Europe, which
now form the elevated plains of central Germany.
But these results are not confined to physiology.
They prove also the ancient condition of the seas
and atmosphere, and the relation of both these
media to light.   For in those remote epochs, the
marine animals were furnished with instruments of
vision, in which the minute optical adaptations were
the same as those which now impart the perception
of light to the living crustacea.   The mutual relations of light to the eye, and of the eye to light,
were, therefore, the same at the time when crustacea first existed in the bottom of the primeval
seas, as at the present moment."*

47. Insects of the Coal Formation. —
Several species of beetle (*curculio*) have been found
in the ironstone of Coalbrook dale, and are figured
by Dr. Buckland; and my museum contains a wing
of a neuropterous insect, closely resembling that of
the living *corydalis* of Carolina ; it is much larger
than the wing of the largest dragon-fly.   This specimen is also from Coalbrook dale ; I discovered it,
together with the limuli on the table, in nodules of
ironstone, for which I am indebted to John Pritchard,
Esq. of Brosely. Not only are the remains of insects
imbedded in the coal strata, but also those of animals,

* Bridgewater Essay.

Q Q

to which they served as food.   Dr. Buckland describes a fossil *scorpion*, discovered by Count Sternberg, in carboniferous argillaceous schist, at Chomle,
S.W. of Prague, in Bohemia.*   This fossil is about
two inches and a half long, and is imbedded in
coal shale, with leaves and fruits.   The legs, claws,
jaws and teeth, skin, hairs, and even portions of
the trachea, or breathing apparatus, are preserved.
It has twelve eyes, and all the sockets remain ; one
of the small eyes, and the left large eye retain their
form, and have the cornea, or outer skin, preserved
in a corrugated or shrivelled state.   The horny
covering remains ; it is neither carbonized nor
decomposed, the peculiar substance of which it is
composed, *elytrine*, having resisted decomposition
and mineralization.

48. FISHES OF THE CARBONIFEROUS SYSTEM.—
Although the remains of fishes are of comparatively
rare occurrence in these formations, yet ten or
twelve species, belonging to extinct and very
remarkable genera, have been established by M.
Agassiz.   In the red sandstone, beneath the coal-
series in Scotland, scales and other remains of fishes
were discovered many years ago, by Dr. Fleming,
by whose kindness the specimens on the table
were added to my museum.   These belong to an
extraordinary fish, named Cephalaspis (*buckler-
headed*) by M. Agassiz, from the head being
covered by a shield, and the bones united into one

* See Dr. Buckland's Bridgewater Essay, pl. 46, p. 406, *et seq.*

osseous case. The scales form elevated bands, and the rays of the fins are covered by the membrane which elsewhere surrounds them. In form, this fish bears a general resemblance to the elongated trilobites of the transition rocks; it is confined to the old red sandstone of England and Scotland.* Another remarkable group of fishes, of this formation, is the *Sauroid*, of which several gigantic species have been found by Dr. Hibbert, in the Burdie-House strata. The teeth of these fishes are large, striated, hollow cones, bearing considerable resemblance to the teeth of crocodiles, with which they were formerly confounded. The scales are thick and strong. The tail, as in all, or nearly all, the fishes found in strata below the magnesian limestone, is unequally lobed, the vertebral column extending to the extremity of the upper lobe, as in the dog-fish of our coast. There are other peculiarities of structure in these fishes, but which my limits will not allow me to discuss; I will only add, that my friend Mr. Murchison will add greatly to our knowledge of the fishes of this epoch in his forthcoming splendid work on the geology of the border counties of England and Wales.

49. RETROSPECT—THE FLORAS OF THE ANCIENT WORLD.—In conclusion, let us review the botanical character of the geological eras, as established by the fossil plants hitherto discovered.

* Mr. Lyell has figured a species of Cephalaspis in vol. iv. Principles of Geology, p. 296.

In the transition rocks, hereafter to be described, about thirteen species of cryptogamic plants, four of which are sea-weeds (*algæ*) and the remainder ferns,

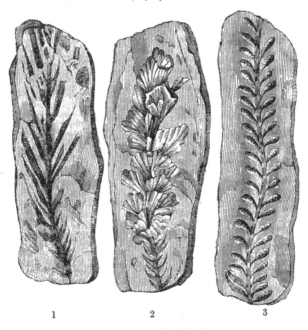

1   2   3

TAB. 73.—PLANTS OF THE NEW RED SANDSTONE.

(Drawn by Miss Ellen Duppa.)

Fig. 1. *Voltzia longifolia.*   2. *Voltzia brevifolia, with the fructification.*
3. *Filicites Scolopendroides.*

comprise all that is known of the vegetable kingdom, anterior to the carboniferous system; no trace remaining of any additional tribes, which may have existed at that period.*

The carboniferous era, as we have seen, abounded in the vascular cryptogamia to a degree unexampled at the present time; the plants belong to species and genera now extinct, but allied to existing types by common principles of organization. The numerical preponderance of the cryptogamia in the coal is such, that while in the present order of nature, they are to the whole number of known plants as one to thirty, at that epoch they were in the proportion of twenty-five to thirty. In the saliferous system, about fifty species have been ascertained, some of which differ from any observed in the coal-measures. They were discovered by M. Voltz, of Strasburgh (to whom I am indebted for the specimens before us), in the shale, or indurated marl, belonging to the new red sandstone at Sulz-les-bains. These fossils consist of a large species of equisetum, four species of ferns, and the frond of a plant bearing some resemblance to the common adder's tongue (*Scolopendrium*) so common on the banks of our woods and hedge-rows. This specimen (Tab. 73, fig. 3) exhibits the back of the leaf, with the fructification. The other vegetables belong to a new genus of coniferæ, named

* Fossil fuci abound in the transition rocks of the Alleghany Mountains, sometimes forming entire layers, one hundred of which occur in a thickness of twenty feet. Dr. Harlan's Medical and Physical Researches, p. 399.

Voltzia, from the discoverer. (Tab. 73. figs. 1, 2.)
They approach the araucaria, or Norfolk Island
pine.* Six species of fuci have been noticed in
this formation. Here, for the first time, appear the
plants allied to the cycadeæ.

In the oolite and lias we find coniferæ, liliaceæ,
palms, ferns, and cycadeæ. The latter hold an
intermediate place between the palms, ferns, and
coniferæ. Some species are very short, as the *zamia ;*
others, thirty feet high, as the *cycas circinnatis.†*
Leaves of cycadeæ are of frequent occurrence in
the shale of the oolite near Scarborough, and several species have been obtained from the Stonesfield
slate. The coniferæ of this epoch exhibit a condensation of the outer margin of each woody layer,
denoting an increase of cold at the latter part of
the autumnal season. In the oolite, the stems and
fruit of a species of pandanus have been discovered. The pandanus, or screw-pine, so named
from the spiral arrangement of its leaves, abounds
in the islands of the Pacific Ocean, and, with the
cocoa-nut, is generally the first vegetable colonist
of the coral islands (Tab. 69, fig. 3). The fossil
fruit was found in the inferior oolite, near Char-
mouth, and is now in Dr. Buckland's collection.
It is of the size of a large orange, and is covered

    * Essai d'une Flore du Grès bigarré, par M. Adolphe Brong-
niart.

    † A *zamia* is introduced in the foreground of the Frontispiece.
It somewhat resembles a pine-apple, with a tuft of leaves on the
apex.

by a stellated rind, or epicarpium, composed of hexagonal tubercles, forming the summits of cells which occupy the entire surface of the fruit.* The fossil plants discovered in the oolitic system comprise four marine species, thirty-nine terrestrial cryptogamia, and forty belonging to coniferæ and other tribes.

The Wealden contains abundance of coniferæ and plants allied to the cycadeæ. One of its ancient forests preserved in stone, and in which the trees still occupy their original position, has been already submitted to our notice (page 335). Equiseta, ferns, cycadeæ, and plants allied to the palms, dracæna, thuya, and yucca, occur in Tilgate forest; and at Hoer, in Scania, M. Adolphe Brongniart has discovered a similar flora.

In the chalk we have the cones and stems of coniferæ, and of plants allied to the yucca. A stem discovered by Mr. Bensted in the Iguanodon quarry near Maidstone, appears related to the Sternbergia of the coal measures; the annular markings of the leaves resemble those of the yucca, or dracæna. Twelve species of fucus, two of conferva, and four of zostera, have been found in the chalk. Dicotyledonous wood, bored by the teredo and fistulana, and water-worn, is common in the line of junction between the galt and green sand. The distinctive character of the flora of the upper secondary formations is the prevalence of cycadeæ.

The tertiary deposites abound in palms and tree-

* Dr. Buckland's Bridgewater Essay, p. 503.

ferns; and dicotyledonous trees (see page 543) begin
to prevail to a greater extent than in the secondary;
these strata include many terrestrial, lacustrine, and
marine plants.   Fossil fruits of existing genera, as
pandanus, cocos, pinus, ulmus, acer, salix, &c., present the essential characters of the modern flora.   The
local accumulation of tropical plants and fruits in
cold and temperate climates has been alluded to in
the previous lectures, and is in accordance with the
difference observable in the zoology of that geological
period.*

In the newer tertiary are imbedded remains of
species, still living in the countries where the
deposites occur.   The fossil foxes and turtles of
Œningen (page 233) were buried amidst beds of
poplars, willows, maples, linden trees, and elms.†
The brown coal of the Rhine (page 251) is composed of similar plants.   In the deposites in actual
progress, the most delicate vegetable remains are
preserved; and in the lacustrine marls of Scotland,
the leaves and seed-vessels of the charæ are found in
a state of fossilization, scarcely distinguishable from
the gyrogonites of the tertiary strata of the Paris basin.

From this review of the botanical epochs which
the present state of geological inquiry enables us to

* The paragraph 1, Lecture V., is, by a typographical error,
headed Fauna and *Flora* of the Chalk.   It will be seen by the
text, that the examination of the fossil plants was not intended,
and would have been foreign to the argument.

† See an interesting account of the fossil plants of Œningen,
by Professor Braun, of Carlsruhe; Dr. Buckland's Essay, p. 510.

establish, we perceive that, from the earliest era, the sea has abounded in the same types of vegetation ; that ferns and other cryptogamia, palms, and coniferæ, have also existed through periods of indefinite duration to the present time : the important difference in the ancient and modern floras being the numerical preponderance of the cryptogamia in the former, and of the dicotyledonous tribes in the latter. The theory of the progressive development of creation, as Dr. Lindley has emphatically remarked, receives no support from the state of vegetation in the early geological epochs. We have no fungi, lichens, hepaticæ, or mosses, in the coal ; but coniferæ, and the most perfectly organized of the cryptogamic class.

The *absence* of other types of vegetation in the transition rocks must not however be received as proof that the flora of that epoch was thus sterile: the only legitimate inference, in the present state of our knowledge, is that the circumstances, under which those strata were accumulated, were unfavourable to the envelopement or preservation of terrestrial plants. We have seen that the existing fundamental distinctions of vegetable structure prevailed also in the earliest secondary formations, a fact in accordance with what we observed in the animal kingdom: and this same unity of purpose and design is manifest in all the varied forms of organization that lived on our planet, through the vast range of time which geological investigations have enabled us to scan.

LECTURE VIII.

1. INTRODUCTION.—In the former lecture the flora of the ancient world constituted the principal object of our investigations; we examined the primeval forests of coniferæ, and the groves of palms and arborescent ferns, which clothed the surface of the soil in that remote epoch of the earth's physical history.   The insects which fluttered among the

tropical vegetation of that early Polynesia, and the fishes and crustacea which abounded in the seas and rivers, were brought in review before us, and we contemplated their extraordinary forms and organization, as preserved by those natural processes

> " Which turned the ocean-bed to rock,
> And changed its myriad living swarms
> To the marble's veined forms." *

The wonderful transmutation into stone of the most delicate structures both of animals and vegetables; the mineralization of the complex visual apparatus of that ancient family of crustacea, the trilobites; and the evidence thus afforded of the actual condition and relation of the waters and of the atmosphere with light, at the period when the ocean swarmed with those singular beings, were considered and explained.

We now advance another stage in our eventful progress; and again we have to examine deposites which have been accumulating for countless ages in the basins of seas fed by rivers and streams, bearing with them the detritus of the strata over which they flowed, and imbedding the remains of the plants and animals that existed at the epoch of their formation. Again we shall find new forms of existence presented to our notice; differing from, but bearing an analogy to the inhabitants of the waters which deposited the marine strata of the

* Mrs. Howitt.

most ancient beds hitherto examined; but altogether dissimilar to those of more modern eras. In vain may we seek for the remains of the mammalia of the tertiary period—of the mollusca, fishes, and reptiles of the chalk—of the colossal oviparous quadrupeds of the country of the Iguanodon—of the dragon-forms of the Jura limestone—of the fish-like lizards of the lias—or of the tropical forests of the coal-measures—all have disappeared; and as the traveller who ascends to the regions of eternal snow, gradually loses sight of the abodes of man, and of the groves and forests, till he arrives at sterile plains, where a few stinted shrubs alone meet his eye; and as he advances, even these are lost, and mosses and lichens remain the only vestiges of organic life; these too at length pass away, and he enters the confines of the inorganic kingdom of nature. In like manner the geologist who penetrates the secret recesses of the globe, perceives at every step of his progress the existing forms of animals and vegetables gradually disappear, while the shades of other creations teem around him; these in their turn vanish from his sight—other new and strange modifications of organic structure supply their place; these also fade away—traces of animal and vegetable life become less and less manifest, till they altogether disappear; and he descends to the primary rocks, where all evidence of organization is lost, and the granite, like a pall thrown over the relics of a former world, conceals for ever the earlier scenes of the earth's physical drama.

2. THE SILURIAN SYSTEM.—By a reference to the chronological arrangement (page 180, and pl. 3), it will be seen that arenaceous and argillaceous strata, with limestone, and an immense thickness of slate rocks, fill up the interval between the lowest member of the carboniferous system, the old red sandstone, and the mica schist, which is the upper-most of the metamorphic or igneous crystalline rocks. The general term of *Transition* was formerly applied to these formations (page 13), and also that of grau, or greywacke.* The upper series is now designated the Silurian system ; † a name selected by Mr. Murchison, by whose able and indefatigable researches the true position and relation of this group of rocks have been determined. This system of deposites is well developed in the border coun-ties of England and Wales, and spreads over a considerable area of South Wales, forming a link which connects the carboniferous series with the ancient slate rocks of that country. The strata are named and characterised by Mr. Murchison accord-ing to the following arrangement :—

* From the German *grau*, grey, and *wacké*, a term employed by the miners to denote a hardened conglomerate.

† Silurian—derived from the *Silures*, the ancient Britons who inhabited the country where these strata are most distinctly developed.

SILURIAN SYSTEM. (Pl. 5, iv.)

(Commencing with the uppermost.)

Upper Silurian.
> *Ludlow rocks* — slightly micaceous grey-coloured sandstone. Blue and grey argillaceous limestone. Dark-coloured shales and flag-stones, with concretions of earthy limestone, containing marine shells, orthoceræ, spiriferæ, and trilobites. Fishes.
> *Wenlock, or Dudley limestone*—sub-crystalline blue and grey limestone—abounding in trilobites, crinoidea, polyparia, spiriferæ, orthoceræ, &c.

Lower Silurian.
> *Wenlock shale*—dark-grey argillaceous shale, with nodules of sandstone.
> *Caradoc sandstone*—shelly limestones, and finely laminated, slightly micaceous, greenish sandstone.
> *Llandeilo flags and limestone.* Freestone, conglomeritic grits, and limestones. Dark-coloured flags. Beds of schist with abundance of trilobites.

Total thickness, nearly 8000 feet.

These beds are entirely of marine origin, and the limestone at Dudley and other places swarms with trilobites, crinoidea, corals, spiriferæ, productæ, and other fossils, with which our previous examination of the carboniferous limestone has made us familiar. The subdivisions introduced are locally important ; but a general analogy prevails throughout the rocks, and there does not appear to be any essential variation in the forms or conditions of organic life, as deducible from the fossils, from the commencement

to the termination of the series. Mr. Murchison
has ascertained that a few peculiar species occur in
each of the divisions; but until the publication of
his splendid work on the Geology of the Frontiers
of England and Wales, our information is too imperfect to warrant farther remark. It will suffice
for our present purpose to learn that beneath the
carboniferous strata there is an immense marine
formation, made up of sands, sandstones, conglomerates, and limestones, some of which were once
coral reefs, while others are wholly formed of crustacean and testaceous exuviæ, belonging to extinct
genera and species, but many of which are similar
to those of the coal. These slabs of Dudley limestone, collected by G. E. Woodhouse, Esq., are, as
you perceive, wholly composed of trilobites, corals,
and shells, held together by indurated clay or mud;
the strata from whence they were obtained, appear
to have been formed in a similar manner to our coral
banks.

3. CUMBRIAN, or CAMBRIAN SLATE SYSTEM.
Pl. 3, fig. 9; Pl. 5, fig. 11. — The deposites I
have thus briefly surveyed, repose on an immense
thickness of strata, generally of a slaty character, and having no distinct assemblage of organic
remains. These rocks extend over a great part
of Cumberland, Westmoreland, and Lancashire,
reaching to elevations of 3000 feet, and giving
rise to the grand scenery of the lakes, and of North
Wales. Beds of this system flank the chain of the

Grampians, and the range of Lammermuir, and occur in Argyleshire, and in the west of Scotland. In Ireland, slate borders the region of primary rocks. Wales may be described as a grand slate formation, with a considerable expansion of greywacke. The slate district of Cornwall is well known. Charnwood Forest, part of Anglesea, and the Isle of Man, are formed of beds belonging to this series. The following tabular arrangement explains the relative position and characters of the subdivisions adopted by Professor Sedgwick :—

CUMBRIAN, OR SLATE SYSTEM, OF NORTH WALES.
(*Commencing with the uppermost.*)

Plynlymmon rocks. Grauwacke and slate, with beds of conglomerates. Thickness, several thousand yards.

Bala limestone. Dark limestone, associated with slate, containing a few species of shells and corals.

Snowdon rocks. Slates, fine-grained, and of various shades of purple, blue, and green. Fine and coarse grauwacke and conglomerate. A few organic remains. Thickness, probably several thousand yards.

The upper dark-coloured schists contain a few corals and fuci ; and Professor Phillips has discovered in the strata of Snowdon two species of corals (*cyathophilla*), and six of shells (*brachiopoda*), of the ancient family of terebratula. The fineness of grain, hardness, colour, and structure of these rocks, are too well known from the universal employment of slate for economical purposes, to require description. The beds termed greywacke slate, contain fragments and pebbles of clay slate, and other rocks,

and have evidently been indurated by high temperature: in proportion as the extraneous substances are
large or small, abundant or scanty, the compound
rock is termed a greywacke conglomerate, or greywacke slate. As Mr. Bakewell observes, if the red
colour were absent in the conglomerates of the old red
sandstone (page 534), those beds would be in every
respect identical with the greywacke of these lower
formations. I may add, that many of the brecciated
clays and shales of the Wealden, if indurated, would
also closely resemble this conglomerate. It would be
a waste of time in a popular discourse to enter upon
a description of the varied lithological characters
or of the geographical distribution of these rocks.
They almost universally occur on the flanks of the
primary rocks; rising up into the most lofty mountain chains, and dipping beneath the newer secondary deposites.   This section, by Mr. Conybeare
(Pl. 6), from the Irish Sea, through Cumberland, by
Durham, to the North Sea, will serve to illustrate
the above remarks ; it shows the elevation of Skiddaw and Saddleback, peaks 3000 feet high, by a
central mass of granite; and the disruptions of
the secondary strata by the intrusion of primary
rocks.   The relative position of the different mem-
bers of the Cumbrian group is seen in this section
(Pl. 5, fig. 2).

In some of the slate rocks of Wales, the laminæ
are so charged with a species of asaphus (page
588), that millions must be imbedded in those

R R

rocks :* in Normandy and Germany, similar remains
are not less abundant; my collection contains specimens presented by MM. Cuvier and Brongniart.
In North America, the slate system of rocks extends
over immense areas; a few orthoceratites and spiriferæ have been found in some of the beds.

4. ORGANIC REMAINS OF THE SILURIAN AND
CUMBRIAN SYSTEMS.—The seas, which deposited
the upper group of limestone, appear, as I have
previously stated, to have swarmed with zoophytes,
crinoidea, brachiopodous mollusca, and trilobites; of
the latter, between fifty and sixty species have been
discovered; some of the limestone beds and schists
are almost wholly composed of their remains.   Mr.
Murchison observes, that no vegetables, except imperfect traces of fuci, have been observed by him
or Professor Sedgwick in any of the deposites below
the old red sandstone; nor any coaly matter, with
the exception of small nests of anthracite.   From
the confusion that prevailed in the classification of
these deposites previously to the labours of Messrs.
Sedgwick and Murchison, the foreign geological
localities of organic remains cannot be fully relied
upon.   In the lower slate system, fossils are of rare
occurrence; and in its few species of fuci, corals,
and shells, we see the last trace of organization, and
arrive at the extreme limits of the animal and vegetable kingdoms of the ancient world.

* De la Beche.

5. METAMORPHIC CHARACTER OF SLATE AND
GREYWACKE —The sedimentary nature of the Silurian system, is too obvious to admit of question;
layers of shells, corals, crustacea, with a few remains
of fishes, imbedded in mud, clay, and sand, together
with coarse water-worn materials, at once evince the
origin and mode of formation of the strata.   When
dykes of basalt or trap traverse or intersect the
limestones or shales, we find them indurated, and
sometimes altogether changed in their lithological
characters.   In the slates, the lines of stratification
are commonly less manifest, and the rocks have a
cleavage, that is, a tendency to split in directions
which have no relation to the lines of deposition, but
have clearly been induced from exposure to a high
temperature, by which the character and arrangement of the constituent substance of the rock have
been altered; for a tendency to a similar structure is
observable when argillaceous beds are found in contact with lavas.   It is also observed that where slate
rocks have been exposed to a still greater degree of
igneous action, the metamorphosis is more complete;
as, for instance, when granite has been erupted in
a state of fusion into fissures and veins of schist.
The greywacke is clearly an indurated conglomerate
of water-worn materials.

The numerous metalliferous veins which occur
in the slate rocks, are either fissures into which
mineral matter has sublimed, or cavities which
appear to have been formed in the rock itself, and

into which the metal has been introduced by segregation.

6. THE METAMORPHIC, or PRIMARY ROCKS.
(Pl. 3, figs. 10—20 ; Pl. 5. fig. II.)—We have at
length passed the boundary which separates the
animate from the inanimate world, and have entered
upon those regions of geological research, in which
all traces of organized beings are lost.

The primary (see pp. 13, 180) or metamorphic
rocks, so called from the supposition that they have
been changed or metamorphosed by igneous agency
since their original formation, are divided into two
natural groups.   1st. Those rocks which, although
of a crystalline structure, and destitute of organic
remains, yet exhibit traces of stratification, and consequently must originally have been formed by sedimentary deposition; and 2dly, those which present
no appearance of regular arrangement, but occur in
amorphous, or shapeless masses, or as dykes and
veins filling up fissures in pre-existing rocks, or
interpolated between regular strata.

7. MICA SCHIST AND GNEISS.—The stratified
metamorphic rocks consist essentially of two groups.
The first, or uppermost, is *Mica Schist,* a slaty
rock, abounding in a mineral called mica (from its
glittering appearance), and quartz, a substance with
which you are acquainted in the form of rock crystals, and of semidiaphanous pebbles, common in
most beds of shingle or beach.   These two minerals are disposed in alternate layers, forming

laminated beds, which are extremely contorted and undulated. The upper divisions of the series bear a considerable resemblance to the argillaceous schists; the lower are of a more quartzose character, probably from having been subject to a greater degree of igneous action.

The *Gneiss* (a German mining term) system, consists of contorted and laminated beds of quartz, felspar, and mica, irregularly stratified; they are in truth stratified granite, for the same substances enter into their composition, as prevail in the amorphous masses of that rock. The gneiss is often found associated and alternating with mica schist, quartz-rock, clay-slate, and a very hard granular rock, called primary limestone. The whole series of stratified metamorphic beds may therefore be considered as partaking of one general lithological character, and with the exception of the calcareous rocks, may have originated from the disintegration and subsquent consolidation of more ancient primary masses. There are various substances associated with this group, as steatite, hornblend schist, chlorite schist, and the beautiful mottled magnesian rock called *serpentine ;* but I must refer you to the elementary works on Geology and Mineralogy, previously cited (page 171).

Mica schist and gneiss are widely spread over and around the unstratified masses of primary rocks of which I shall presently treat. They are scarcely known in England, but are the prevailing rocks in

the Highlands of Scotland, in the Hebrides, and in
the mountain ranges of Ireland; they also occur in
Skiddaw, and on Snowdon in Wales. Their geographical distribution over Europe and America is of
vast extent. This system abounds in metalliferous
veins.

The position of the gneiss, in relation to granite
and the slate rocks, is seen in this section, Pl. 5,
fig. II.

8. UNSTRATIFIED METAMORPHIC ROCKS—
GRANITE.   (Pl. 3, fig. 18; Pl. 5, fig. II. Pl. 6).—
The unstratified crystalline rock, Granite (so named,
from its granular structure), constitutes at once the
foundation upon which all the strata of which we
have spoken are spread out, and the great frame-
work of the earth's surface, rising to the loftiest
heights, and stretching into mountain chains, which
mark the grand, natural divisions of the physical
geography of the globe.

Although presenting many varieties in the proportion and colour of its ingredients, granite is
essentially composed of three substances, which are
easily recognised in the blocks of which our pave-
ments and bridges, and other works, are constructed.
These are *mica*, known by its silvery or black glittering aspect; quartz, by its grey glassy appearance;
and felspar, which forms the opaque white, pink, or
yellowish masses, oftentimes seen in sections, as
long angular crystals, which from their size and colour constitute a striking feature in the stones. In

some granite rocks, talc and hornblende occur, and the mica is wanting; these are called syenite: the masses composed of crystals of felspar, in a base of earthy felspar, constitute porphyry. Granite is found almost everywhere beneath gneiss and mica schist, and in contact with rocks of all subsequent periods, rising into enormous masses and peaks ; in the British islands it occurs in Cornwall, Dartmoor, Skiddaw, and Shapfell in Cumberland, Glen Tilt, Ben Nevis, &c. It is also found in veins which traverse not only other rocks, but also masses of granite, thus proving periodical formations of this crystalline rock.

In some instances, a tendency to a columnar or prismatic arrangement, is observable in granite ; and the granitic porphyry of Corsica (*Napoleonite*) presents an orbicular structure, in which balls or spheroids of concentric and alternate coats of hornblende and compact felspar are disseminated throughout the mass, with much regularity. An instance of the elevation of the superincumbent beds of slate by granite, is seen in this section of the Cumberland rocks (Pl. 6); and the distribution of the primary rocks in England is shown in the map (Pl. 6).

9. VOLCANIC AGENCY. — Throughout the vast series of stratified rocks the effect of water was everywhere apparent ; and the existence of dry land, streams, rivers, seas, animals, and plants, unequivocally manifest ; and, although the influence of high temperature was seen in the

altered character of rocks in contact with ancient
lava currents, yet its effects were comparatively
but feebly displayed. The metamorphic rocks, on
the contrary, present the most unquestionable proofs
of their igneous origin; and many can scarcely be
distinguished from the products of modern volcanoes.
To unveil the mystery in which their origin is in-
volved, we must, therefore, as in our previous inquiry,
examine those natural operations which are producing analogous results; and I, therefore, purpose in
this place to review the phenomena presented by
existing volcanoes.

Volcanic action is defined by Humboldt to be the
influence exercised by the internal heat of a planet on
its external surface, during its different states of refrigeration; by which concussions of the land, or earthquakes, and the elevation and subsidence of large
portions of the solid crust, are produced. The number of existing volcanoes is estimated at about 200,
of which 116 are situated in America, or its islands.
In the previous discourses, many of the effects of
igneous agency were noticed, namely, the subsidence and elevation of the Temple of Serapis (page
83); the gradual rise of Scandinavia (page 92);
the upheaving of the sea-coast of Chili (page 87);
and other changes of a like nature. As we suc-
cessively examined the tertiary and secondary
formations, proofs of similar phenomena, in every
geological epoch, were equally manifest; the foci of
volcanic action were found to have varied, but

throughout the cycle of physical changes contemplated by geology, the volcano and the earthquake appeared to have been in active operation.

The present grand European centre of volcanic action is in southern Italy, which has for ages been in a state of energy ; Etna, Vesuvius, and the Lipari isles, being the vents through which its incandescent materials have escaped. The action of its fires on the calcareous rocks of the Appennines evolved the carbonic acid of the waters, which deposited the travertine of Pæstum, Solfatara, &c. previously described (page 52).*

10. VESUVIUS. — The celebrated mountain of Vesuvius, or Somma, is about four thousand feet high, and its summit is now broken and irregular ; but, observes Mr. Lyell,† when northern Italy was first colonized by the Greeks, "its cone was of a regular form, with a flattish summit, where the remains of an ancient crater, nearly filled up, had left a slight depression, covered in its interior by wild vines, and with a sterile plain at the bottom." From the earliest period to which tradition refers, to the first century of the christian era, the mountain had exhibited no appearance of activity, but we then arrive at a crisis in the volcanic action of this district, which gave rise to " one of the most interesting events witnessed by man during the brief period

* Consult Mr. Lyell's admirable description of modern volcanoes and their effects. Principles of Geology, vol. ii.

† Principles of Geology, vol. ii. p. 67, *et seq.*

throughout which he has observed the physical
changes of the earth's surface." In the year 63 after
Christ, the volcano exhibited the first symptom of
internal change, in an earthquake which occasioned
considerable damage to many neighbouring cities,
and of whose effects traces may yet be witnessed
among the interesting memorials of the awful cata-
strophe which soon afterwards took place.* After this
event, slight shocks of earthquakes were frequent,
when on the 24th of August, in the year 79, a
tremendous outburst of the long pent-up incandescent materials of the volcano took place, and spread
destruction over the surrounding country, overwhelming three cities, with many of their inhabitants, and burying all traces of their existence beneath immense accumulations of ashes, sand, and
scoriæ. All the fearful circumstances connected
with this event, and the attendant physical phenomena, are so well known, that it is unnecessary to
dwell upon the subject.

From that period the internal fires of Italy have
resumed their ancient focus, and Vesuvius, with
occasional periods of tranquillity, has been more or
less active to the present time. The principal eruptions are recorded in Mr. Lyell's interesting volume.
I will only mention one remarkable event which
happened in 1538. After frequent earthquakes, a
gulf opened near the town of Tripergola, which
discharged mud, pumice-stones, and ashes, and

* Daubeny on Volcanoes, p. 152. Scrope on Volcanoes.

threw up in one day and night a mound of volcanic
materials, now called *Monte Nuovo*, a mile and a
half in circumference at the base, and 440 feet in
height; at the same time the coast to beyond
Puzzuoli was permanently elevated many feet above
the level of the Mediterranean.

11. MODERN ERUPTIONS OF VESUVIUS.—In the
early periods of activity, violent explosions, with
showers of scoriæ, ashes, and sand, characterised the
eruptions of Vesuvius;* but since the existence of
the present crater, lava-currents have generally
been ejected.   The appearance of an ordinary
eruption, seen by night, is thus graphically described
by a late traveller:—

" It was about half-past ten when we reached the foot of the
craters, which were both tremendously agitated; the great vent
threw up immense columns of fire, mingled with the blackest
smoke and sand.   Each explosion of fire was preceded by a
bellowing of thunder in the mountain. The smaller mouth was
much more active; and the explosions followed each other so
rapidly, that we could not count three seconds between them.
The stones which were emitted were fourteen seconds in falling
back to the crater; consequently, there were always five or six
explosions—sometimes more than *twenty*—in the air at once.
These stones were thrown up perpendicularly, in the shape of
a wide-spreading sheaf, producing the most magnificent effect
imaginable.   The smallest stones appeared to be of the size of
cannon-balls; the greater were like bomb-shells; but others
were pieces of rock, five or six cubic feet in size, and some of
most enormous dimensions: the latter generally fell on the
ridge of the crater, and rolled down its sides, splitting into

* The craters of Auvergne, that exhibit no traces of lava
currents, are also supposed to have been produced by explosions.

fragments as they struck against the hard and cutting masses
of cold lava.   The smoke emitted by the smaller cone was white,
and its appearance inconceivably grand and beautiful ; but the
other crater, though less active, was much more terrible ; and the
thick blackness of its gigantic volumes of smoke partly concealed
the fire which it vomited.   Occasionally both burst forth at the
same instant, and with the most tremendous fury ; sometimes
mingling their ejected stones.

" If any person could accurately fancy the effect of 500,000
sky-rockets darting up at once to a height of three or four
thousand feet, and then falling back in the shape of red-hot
balls, shells, and large rocks of fire, he might have an idea of
a single explosion of this burning mountain ; but it is doubtful
whether any imagination can conceive the effect of one hundred
of such explosions in the space of five minutes, or of twelve hundred or more in the course of an hour, as we saw them !   Yet
this was only a part of the sublime spectacle before us.

" On emerging from the darkness, occasioned by the smaller
crater being hidden by the large one, as we passed round to the
other side of the mountain, we found the whole scene illuminated by the river of lava, which gushed out of the valley formed
by the craters and the hill on which we now stood. The fiery current was narrow at its source, apparently not more than eighteen
inches in breadth; but it quickly widened, and soon divided
into two streams, one of which was at least forty feet wide, and
the other somewhat less : between them was a sort of island,
below which they reunited into one broad river, which was at
length lost sight of in the deep windings and ravines of the
mountain."[*]

In an eruption witnessed by Sir W. Hamilton,
jets of liquid lava, mingled with stones and scoriæ,
were thrown up to a height of ten thousand feet.
The liquid streams of lava issue with great velocity,
and are in a state of perfect fusion ; but as they

* From the Saturday Magazine.

cool on the surface, they crack, and the matter
becomes vesicular, or porous ; at a considerable
distance from their source, they resemble a heap of
scoriæ, or cinders, from an iron foundry, rolling
slowly along, and falling with a rattling noise, one
over the other.

12. VOLCANIC PRODUCTS OF VESUVIUS. — The
cone of Vesuvius consists of concentric coatings of
lava, sand, and scoriæ, inclining outwards from the
axis of the mountain in an angle of from 30º to 45º :
a section would exhibit the structure here represented (Pl. 4, fig. 1).   The fissures and rents produced in the cooled lavas and beds of volcanic
products, by the earthquakes which generally precede
eruptions, become filled up by subsequent ejections
of melted matter, and form dykes and veins ; when
these are injected into masses of materials which
readily decompose, the solid and durable matter of
the dyke remains in the form of vertical walls, of
which many striking examples occur in Etna, and
are figured and described by Mr. Lyell.*

Lava is a term applied to any rock liquefied by
heat ; when consolidated by cooling, it may be in a
state of scoria, pumice, basalt, obsidian, trachyte, &c.
according to its mineral composition, and its slow or
rapid refrigeration.   The chief constituents of lavas
are the substances called felspar and augite, and titaniferous iron, and the lavas are classed according
as either of these ingredients predominates.   When

* See Principles of Geology, vol. iii. figs. 102, 105.

the felspar prevails, the mass is called *trachyte*, which is generally of a coarse grain, with a harshness of texture, and a degree of porosity; when the grain is fine and compact, but irregular, it constitutes *trachytic porphyry;* when the particles are so fused as to have a resinous or glassy texture, it forms *pitchstone* and *obsidian*. If augite or titaniferous iron constitute a large proportion of a rock, it is termed *basalt;* when the structure is slaty, it forms *clinkstone*.

The Vesuvian lavas present considerable variety of appearance and structure; pumice stone, scoriæ, and vesicular, or full of hollow cells; compact and heavy like iron; yellowish, or greenish-grey, or black; and internally spotted with red, yellow, brown, or grey; crystallized quartz and hornblende, so abundant in granite, are extremely rare; but mica occurs plentifully in some recent trachytes. Pumice is supposed to have been produced from a considerable disengagement of vapour, while the lava was in a plastic, but not entirely in a fluid, state; the escape of the gaseous matter giving rise to the porous structure of this substance. But I will not embarrass you by naming and describing minerals, the nature of which cannot be thoroughly understood without patient examination of specimens. The number of simple minerals found in the rocks of Vesuvius amount to 400 species; of these my collection contains an extensive and

* Scrope on Volcanoes, p. 85.

valuable series, through the kindness of the Marquis of Northampton.

In some of the ancient lavas of Vesuvius, there are decided indications of a concretional and prismatic structure, and a tendency to divide into columns. Tuff, a term which I have made use of in this discourse, designates beds formed of scoriæ, sand, and ashes, which have either been wafted by the winds, and fallen into the sea, or washed down by torrents on the plains, and agglutinated together. The conglomeration called peperino, and the lapilli or pisolitic globules of earthy matter, appear to have been formed by showers of rain falling through an eruption of fine volcanic sand. With this rapid notice of a few of the principal constituents of the products of volcanic eruptions, I must pass to the consideration of other phenomena connected with this subject.

The effects produced by lavas, and their slow or rapid progress, depend, of course, on their degree of incandescence and fluidity. Lava currents from Vesuvius have flowed a mile and a half in fourteen minutes ; others have reached the sea in three hours from the summit of the mountain, a distance of 3200 yards. The lava stream which destroyed Catania in 1669, was fourteen miles long and five wide. In Etna, currents have been traced forty miles in length ; and a stream that issued from Mount Hecla, in Iceland, is computed at ninety-four miles by fifty.*   Some streams are very sluggish, and

* Scrope on Volcanoes, p. 92.

diverted from their course by any considerable
obstacle; many retain a high temperature for many
years. A curious circumstance occurs when trees
are enveloped by lava : the upper parts and the
branches alone burst out in a flame, while the trunk
is only carbonized ; and, if subsequently removed,
may leave its impression in a hollow, cylindrical
tube within the solid rock. Such moulds are common in the Isle of Bourbon, in those lava currents
that have extended their ravages through forests
of palms.*

13. MOUNT ETNA.—This volcanic cone, which
is entirely composed of lavas, rises majestically to
an altitude of nearly two miles, the circumference
of its base exceeding 180. Compared with this prodigious mass of igneous products, Vesuvius sinks
into insignificance ; for while the lava-streams of
the latter do not exceed seven miles, those of Etna
are from fifteen to thirty miles in length, five in
breadth, and from fifty to one hundred feet in
thickness.† The grand feature of Etna is a zone of
subordinate volcanic mountains, some of which are
covered with forests, while others are bare and arid
like those of Auvergne. The base, for an extent of
twelve miles upwards, is richly cultivated, and
abounds in vineyards and pastures, with towns,
villages, and monasteries. The middle region is
woody, covered with forests of oak and chestnut,

* Scrope on Volcanoes, p. 107.
† Daubeny on Volcanoes.

and a luxuriant vegetation. From about a mile below the summit, all is sterility and desolation, and the highest point is covered with eternal snow. The crater, from which a column of vapour constantly escapes, is about a quarter of a mile high, and three-quarters of a mile in circumference. The varied and picturesque scenery of this extraordinary mountain, the physical changes now in progress, as well as those which have taken place in periods far beyond all human history or tradition, but of which natural records still remain, are sketched by Mr. Lyell with the vigour and fidelity which characterise all the productions of his pen.[*]

14. PHLEGREAN FIELDS AND LIPARI ISLES.— The volcanic district of Puzzuoli and Cumæ, on the bays of Baiæ and Naples, called the Phlegrean Fields, and in which are situated Monte Nuovo, Monte Barbaro, the Solfatara, and the temple of Serapis, of which I have already spoken, presents a series of cones and crateriform basins; some of which contain lakes, as those of Avernus and the Lucrine. These volcanic mounds are formed of felspathic tufa, occasionally containing marine shells and carbonized wood, and are covered by beds of loose tufaceous conglomerate. They are supposed by Mr. Scrope to have been produced by numerous submarine eruptions, each from a fresh focus, on a shallow shore.[†] The Solfatara con-

[*] Principles of Geology, vol. ii.
[†] Scrope on Volcanoes, p. 179.

stantly evolves aqueous vapour, with muriatic and sulphureous exhalations. The celebrated incrusting springs (page 50) derive their properties from the carbonic acid gas, so largely disengaged by subterranean volcanic action on limestone rocks.

The Lipari Isles, between Naples and Sicily, lying, as it were, midway between Vesuvius and Etna, present a character very analogous to the district I have just described. The crater of one of the islands, Stromboli, has been in constant activity from the earliest historical period. I am induced briefly to allude to this group of volcanic mounds, that I may explain the nature of an interesting suite of specimens collected by William Tennant, Esq. from the cliffs of St. Calogero, and now in my museum. These cliffs, which are about two hundred feet high, extend four or five miles along the coast, and consist of horizontal beds of volcanic tuff. From the perennial emanation of sulphureous vapour, the rocks are decomposed; alum, gypsum, and other sulphuric salts, are formed; muriate of ammonia and silky crystals of boracic acid are also among the specimens before you. The dark clays have become yellow, white, red, pink, chequered, and marked with stripes of various colours. Veins of chalcedony and opal occur, and pumice-stone and obsidian are abundant. Dykes and veins of trachyte intersect the tuff in every direction, and bear a striking resemblance to the similar intrusions of trap into the secondary

strata. I will conclude this notice of active volcanoes, by an account of a celebrated burning
mountain in the Pacific.

TAB. 74.—THE VOLCANO OF KIRAUEA, IN HAWAII.
(*From Ellis's Polynesian Researches.*)

15. VOLCANO OF KIRAUEA.—Of the existing
volcanoes, that of Kirauea in Hawaii, (formerly
called Owhyhee,*) so graphically described by
Mr. Stewart† and Mr. Ellis,‡ exhibits volcanic action in the most sublime and imposing aspect. The
whole island, which covers an area of 4000 square
miles, is a complete mass of volcanic matter in

* One of the Sandwich Islands, well known as the scene of
the murder of Capt. Cook.
† Lord Byron's Visit to Hawaii.   ‡ Pol. Res. vol. iv.
s s 2

different states of decomposition, perforated by innumerable craters, and rising to an altitude of
16,000 feet. It is in fact a hollow cone, with numerous vents, over a vast incandescent mass, which
doubtless extends beneath the bed of the ocean ; the
island forming a pyramidal funnel from the furnace
beneath to the atmosphere.   The following account
of a visit to the crater, affords a striking picture of
the splendid but awful spectacle which it presents.

"After travelling over extensive plains, and climbing rugged
steeps, all bearing testimony of volcanic origin, the crater of
Kirauea suddenly burst upon our view.   We found ourselves on
the edge of a steep precipice, with a vast plain before us, fifteen
or sixteen miles in circumference, and sunk from two hundred
to four hundred feet below its original level.   The surface of
this plain was uneven, and strewed over with large stones and
volcanic rocks ; and in the centre of it was the great crater, at
the distance of a mile and a half from the precipice on which
we were standing.   We proceeded to the north end of the ridge,
where, the precipice being less steep, a descent to the plain
below seemed practicable ; but it required the greatest caution,
as the stones and fragments of rock frequently gave way under
our feet, and rolled down from above.   The steep which we had
descended was formed of volcanic matter, apparently of light-red
and grey vesicular lava, lying in horizontal strata varying in
thickness from one to forty feet.   In a few places the different
masses were rent in perpendicular and oblique directions, from
top to bottom, either by earthquakes, or by other violent con-
vulsions of the ground connected with the action of the adjacent
volcano.   After walking some distance over the plain, which in
several places sounded hollow under our feet, we came to the
edge of the great crater.   Before us yawned an immense gulf
in the form of a crescent, about two miles in length from north-
east to south-west, one mile in width, and 800 feet deep.   The

bottom was covered with lava, and the south-west and northern parts were one vast flood of burning matter. Fifty-one conical islands of varied form and size, containing as many craters, rose either round the edge or from the surface of the burning lake. Twenty-two constantly emitted columns of grey smoke, or pyramids of brilliant flame : and at the same time vomited from their ignited mouths streams of lava, which rolled in blazing torrents down their black indented sides into the boiling mass below, (see Tab. 76.) The existence of these conical craters led us to conclude, that the boiling caldron of lava did not form the focus of the volcano ; that this mass of melted lava was comparatively shallow; and that the basin which contained it was separated by a stratum of solid matter from the great volcanic abyss, which constantly poured out its melted contents through these numerous craters, into this upper reservoir. We were farther inclined to this opinion from the vast columns of vapour continually ascending from the chasms, in the vicinity of the sulphur banks and pools of water, for they must have been produced by other fire than that which caused the ebullition in the lava at the bottom of the great crater ; and also by noticing a number of small craters in vigorous action high up the sides of the great gulf, and apparently quite detached from it. The streams of lava which they emitted rolled down into the lake and mingled with the melted mass, which, though thrown up by different apertures, had perhaps been originally fused in one vast furnace. The sides of the gulf before us, although composed of different strata of ancient lava, were perpendicular for about 400 feet, and rose from a wide horizontal ledge of solid black lava, of irregular width, but extending completely round. Beneath this ledge the sides sloped gradually towards the burning lake, which was, as nearly as we could judge, three or four hundred feet lower. It was evident that the large crater had been recently filled with liquid lava up to this black ledge, and had, by some subterranean canal, emptied itself into the sea, or upon the low land on the shore; and in all probability this evacuation had caused the inundation of the Kapapala coast, which took place, as we afterwards learned, about three weeks prior to our visit. The grey, and in some places apparently calcined sides of the great crater before us; the fissures which intersected the surface of the plain on which we were standing; the long banks of sulphur on the opposite sides of the abyss; the vigorous action of the numerous small craters on its borders.; the dense columns of vapour and smoke that rose out of it, at the north and south ends of the plain, together with the ridge of steep rocks by which it was surrounded, rising three or four hundred feet in perpendicular height; presented an immense volcanic panorama, the effect of which was greatly augmented by the constant roaring of the vast furnaces below."

16. STEWART'S VISIT TO KIRAUEA.—In June 1825, Mr. Stewart, accompanied by Lord Byron, and a party from the Blonde frigate, went to Kirauea, and descended to the bottom of the crater.

"The general aspect of the crater," observes Mr. Stewart, "may be compared to that which the Otsego Lake would present, if the ice with which it is covered in winter were suddenly broken up by a heavy storm, and as suddenly frozen again, while large slabs and blocks were still toppling, and dashing, and heaping against each other, with the motion of the waves. At midnight the volcano suddenly began roaring, and labouring with redoubled activity, and the confusion of noises was prodigiously great. The sounds were not fixed or confined to one place, but rolled from one end of the crater to the other; sometimes seeming to be immediately under us, when a sensible tremor of the ground on which we lay took place; and then again rushing on to the farthest end with incalculable velocity. Almost at the same instant a dense column of heavy black smoke was seen rising from the crater directly in front, the subterranean struggle ceased, and immediately after flames burst from a large cone, near which we had been in the morning, and which then appeared to have been long inactive. Red-hot stones, cinders, and ashes, were also propelled to a great height with immense violence; and shortly after the molten lava came boiling up, and flowed down the sides of the cone and over the surrounding scoriæ in most beautiful curved streams, glittering with a brilliancy quite indescribable. At the same time, a whole lake of fire opened in a more distant part. This could not have been less than two miles in circumference, and its action was more horribly sublime than any thing I ever imagined to exist even in the ideal visions of unearthly things. Its surface had all the agitation of an ocean; billow after billow tossed its monstrous bosom in the air; and occasionally those from different directions burst with such violence, as in the concussion to dash the fiery spray forty or fifty feet high. It was at once the most splendid and fearful of spectacles."

17. EARTHQUAKES.—I have indulged in these long extracts, because the vivid pictures which they present of the phenomena attendant on volcanic action cannot fail to produce a powerful impression on the mind, and cause it to revert to those prin-ciples enunciated in the first lecture, which taught us that the early condition of the earth, and of the worlds around us, may have been one of vapour or fluidity (page 22). Here we see the most solid and durable materials of the globe reduced to a liquid state—seas of molten rocks, with their waves and billows, their surge and spray, giving birth to tor-rents and rivers, which, when cooled, become the hardest and most indestructible mineral masses on the surface of our planet!

The constant escape of aeriform fluids from vol-canic vents ; the irresistible force which such elastic vapours exert when pent up and compressed—an

effect with which our steam-boats and locomotive
engines have made every one familiar ; the immense
production of such gaseous elements which must be
taking place in the interior of the globe, from the
igneous action which we have seen is going on un-
remittingly ; afford a satisfactory explanation of the
nature and cause of earthquakes, and of those
elevatory movements by which the foundations of
the deep are broken up, and raised into chains
of mountains thousands of feet above the level of
the sea.   The volcanic vents are, in truth, the
safety-valves from which the caloric from the interior of the earth escapes into the atmosphere :
when these channels become choked up, the confined gases occasion earthquakes, elevations and
dislocations of the crust of the earth, until the
obstruction from the former craters is removed, or
new vents are established.

18. VOLCANIC ISLANDS IN THE MEDITERRA-
NEAN.—These effects take place alike indiscrimi-
nately, either on the land or beneath the waters of
the ocean.   The volcanic foci of southern Italy are
certainly not confined to the land, but extend beneath the bed of the Mediterranean, of which the
appearance of new shoals and islands, affords con-
clusive evidence.   Livy informs us that an event of
this kind, which took place about the period of the
death of Hannibal, together with other volcanic
phenomena, so terrified the Roman people, as to
induce them to decree a supplication to the gods,

to avert the displeasure of heaven, which these
prodigies were supposed to denote.*

In 1831 a volcanic island arose in the Mediterranean, about thirty miles off the S.W. coast of
Sicily. It was preceded by a fountain of steam and
water, and at length a small island gradually appeared, having a crater on the summit, which ejected
scoriæ, ashes, and volumes of vapour; the sea
around was covered with floating cinders and dead
fish. The scoriæ were of a greyish black colour,
as you may observe from this specimen.† The crater
reached an elevation of nearly 200 feet, and with a
circumference of about three miles, having a circular
basin full of boiling water of a dingy red colour.
It continued in activity for three weeks, and
then gradually diminished, so that no trace now
remains of its existence, except in reefs and shoals.
Its appearance, when visited by M. Constant Pre-
vost, is shown in this representation, with which he
favoured me. Mr. Lyell observes, that from the
facts that have been obtained it is certain that a
hill, 800 feet high, was here formed by a submarine
volcanic vent in the course of a few weeks. The occurrence of shoals of dead fish will not fail to remind
you of the ichthyolites of Monte Bolca (page 234):
and we cannot doubt that vast numbers were im-

* " Nuntiatumque erat haud procul Siciliâ insulam quæ
nunquam ante fuerat novam editam e mari esse."—LIVY, lib.
xxxix. c. 56.

† Collected from the island when it had reached its utmost
extent.

bedded in the erupted mineral masses at the bottom
of the Mediterranean; when these shall be elevated
above the waters, and explored by some Agassiz
of future times, the then fossil fish of the Mediterranean may afford interesting subjects for the contemplation of the geologist, and the philosopher.

19. ORGANIC REMAINS IMBEDDED BENEATH
LAVA CURRENTS.—In the course of these inquiries,
we have been familiarized with the striking contrast
presented by the effects of high temperature, exerted under great pressure, to those resulting from
heat and combustion in the open air. Thus we
have seen that in the earliest geological eras,
eruptions of basalt have burst through and overflowed sedimentary strata, and yet the most delicate
animal and vegetable substances have remained;
transmuted, indeed, into stone, but still retaining
their original structure—as, for instance, the vegetables of the carboniferous system, and the shells
and corals of the lias, oolite, and of the chalk. In the
cretaceous formation of Glaris, although the strata
have been converted into slate by igneous agency,
the fishes still remain (page 312)—the limestone of
Monte Bolca, though capped with basalt, yet swarms
with ichthyolites (page 234)—the fiery currents of
Auvergne have flowed over the lacustrine limestones, and still the remains of insects, serpents,
birds, and quadrupeds, are uninjured (page 245)—
the tertiary forests of the Andes, which grow on
beds of lava, and now lie buried beneath volcanic

masses of prodigious thickness, preserve their forms
unaltered (page 258)—and the bones of the dodo
are found imbedded in marlstone, covered by lava
of recent origin (page 108).

20. ICE PRESERVED BY RED-HOT LAVA.—A
circumstance of a very extraordinary nature is
described by Mr. Lyell—that of the preservation
for ages of a glacier, or bed of ice, from having
been covered and protected by a flood of red-hot
lava.*   The extraordinary heat experienced in the
south of Europe, during the summer and autumn
of 1828, caused the usual supplies of ice entirely to
fail.   Great distress was consequently felt for want
of a commodity, regarded in those countries rather
as an article of necessity than of luxury.   Etna was,
therefore, carefully explored, in the hopes of dis-
covering some crevice, or natural grotto on the
mountain, where drift snow was still preserved.
Nor was the search unsuccessful; for a small mass
of perennial ice, at the foot of the highest cone,
was found to be part of a large, continuous glacier,
covered by a lava current.   The ice was quarried,
and the superposition of the lava was ascertained to
continue for several hundred yards; unfortunately,
the ice was so extremely hard, and the excavation so
expensive, that there is no probability of the operations being renewed.   Mr. Lyell explains this apparently paradoxical fact, by supposing that a deep
mass of drift snow was covered by a stream of

* Principles of Geology, vol. ii. p. 124.

volcanic sand, which is an extremely bad conductor
of heat; thus the subsequent liquid lava might have
flowed over the whole without affecting the ice
beneath, which at such a height (ten thousand
feet above the level of the sea) would endure as
long as the snows of Mont Blanc, unless melted by
volcanic heat from below.*

21. HERCULANEUM AND POMPEII.—But all these
phenomena are far surpassed in interest by the wonderful preservation of the cities, which were over-
whelmed by the first recorded eruption of Vesuvius.
In the words of one of the most eloquent and philosophical writers of our times, " After nearly seventeen centuries had rolled away, the city of Pompeii
was disinterred from its silent tomb, all vivid with
undimmed hues; its walls fresh as if painted yes-
terday; not a tint faded on the rich mosaic of its
floors; in its forum the half-finished columns, as left
by the workman's hand; before the trees in its gar-
dens the sacrificial tripod; in its halls the chest of
treasure; in its baths the strigil; in its theatres the
counter of admission; in its saloons the furniture
and the lamp; in its triclinia the fragments of the
last feast; in its cubicula the perfumes and the
rouge of faded beauty; and everywhere the skeletons of those who once moved the springs of that
minute, yet gorgeous machine of luxury and of
life."†

* Principles of Geology, vol. ii. p. 126.
† The Last Days of Pompeii, by E. L. Bulwer, Esq.

The cities of Herculaneum, Pompeii, and Stabiæ, were buried beneath an accumulation of ashes and scoriæ, to a depth of from sixty to one hundred and twenty feet. No traces have been perceived of lava currents or of melted matter; showers of sand, cinders, and scoriæ, with loose fragments of rocks, were the agents of desolation. The various utensils and works of art, as you may observe in the lamps, vases, beads, and instruments in my museum, exhibit no appearance of having suffered by the action of fire. Even the delicate texture of the papyri appears to have sustained more injury from the effects of moisture and exposure to the air than from heat. In Pompeii, the sand and stones are loose and unconsolidated; but in Herculaneum, the houses and works of art are imbedded in solid tuff, which must have originated either from a torrent of mud, or from ashes moistened by water. Hence statues are found unchanged, although surrounded by hard tuff, bearing the impressions of the minutest lines. The beams of the houses have undergone but little alteration, except that they are invested with a black crust. Linen and fishing-nets, loaves of bread with the impress of the baker's name; even fruits, as walnuts, almonds, and chestnuts, are still distinctly recognisable. The remarkable preservation, for nearly 2000 years, of whole cities, with their houses, furniture, and even the most perishable substances, beneath beds of volcanic rocks, may be compared to those geological changes, by which the forests of an earlier world, and the remains of the colossal dragon-forms which inhabited the ancient land and waters, have been perpetuated.

22. SILLIMAN ON THE NATURE OF GEOLOGICAL EVIDENCE.—Although in this stage of our inquiry, no farther exemplification of the nature of geological evidence is, I trust, necessary, yet I cannot deny myself the pleasure of pointing out to you the admirable manner in which Professor Silliman has illustrated the principles of geological induction, by a reference to the discovery of the buried Roman cities.

"When in 1738," he observes, "the workmen, in excavating a well, struck upon the theatre of Herculaneum, which had been buried for seventeen centuries beneath the lava of Vesuvius; when subsequently (1750) Pompeii was disencumbered of its volcanic ashes, and thus two ancient cities were brought to light; had history been as silent respecting their existence, as it was of their destruction, would not all observers say, and have not all actually said—Here are the works of man, his temples, his houses, furniture, and personal ornaments; his very wine and food; his dungeons, with skeletons of the prisoners chained in their awful solitudes, and here and there a victim overtaken by the fiery storm?   Because the soil had formed, and grass and trees had overgrown, and successive generations of men had erected their abodes over the entombed cities, and because these were covered by lava and cinders,—still does any one hesitate to admit that they were once real cities; that they stood upon what was then the surface of the country; that their streets once rang with the noise of business; their halls and theatres with the voice of pleasure; and that they were over-whelmed by the eruptions of Vesuvius, and their place blotted out from the earth and forgotten?   These inferences no one can dispute—all agree in the conclusions to be drawn. When, moreover, the traveller sees the cracks in the walls of the houses of Pompeii, and observes that some of them have been thrown out of the perpendicular, and have been repaired and shored up with props, he infers that the fatal convulsion was not the first, and that these cities must have been shaken to their foundation by the effects of previous earthquakes. In like manner the geologist reasons respecting the physical changes that have taken place on the surface of our globe. The crust of the earth is full of crystals and crystallized rocks; it is replete with the entombed remains of animals and vegetables, from mosses and ferns to entire trees—from the impressions of plants to whole beds of coal. It is stored with the remains of animals, from the minutest shell-fish to the most stupendous reptiles. It is checquered with fragments, from fine sand to enormous blocks of stone. It exhibits in the materials of its solid strata every degree of attrition, from the slightest abrasion of a sharp edge or angle, to the perfect rounding which produces globular and spheroidal forms of exquisite finish. It abounds in dislocations and fractures; with injections and filling up of fissures with foreign rocky matter; with elevations and depressions of strata in every position, from the horizontal to the vertical. It is covered with the wreck and ruins of its former surfaces; and, finally, its ancient fires, although sometimes for a while dormant, have never been wholly extinguished, but still find an exit through volcanic mouths. When we reflect upon these phenomena, we cannot hesitate to infer that the present crust of the earth is the result of the conflicting energies of physical forces, governed by fixed laws; that its changes began from the dawn of creation, and that they will not cease till its materials and its physical laws are annihilated."[*]

23. BASALT, or TRAP.—I return from this digression to the consideration of *Whin, Trap, Basalt,* and *Clinkstone;* terms which designate

[*] Introduction to the third American edition of Bakewell's Geology, edited by Benjamin Silliman, LL.D., &c. 1 vol. 8vo.

different varieties of an ancient volcanic rock, the
nature of which I have already explained. Basalt
occurs sometimes in veins or dykes, which traverse
rocks of all ages, filling up fissures or crevices; and
at others, in layers spread over the surface of the
strata, or interposed between them. In the diagram,
Pl. 3, 15, a trap-dyke is represented traversing the
secondary formations, and underlying the tertiary
(see Pl. 4, figs. vi. vii.). Many modern lavas differ
so little from basalt, that it is unnecessary to adduce
proof of the volcanic nature of this rock. It often
occurs in the form of regular pillars or columns
clustered together; or, in scientific language, has a
columnar structure, a character also observable in
some recent lavas; the columnar basalt of the tertiary
epoch has already been noticed (pages 248—250).
This structure is found by some highly interesting
and philosophical experiments, to have originated
from the manner in which refrigeration took place.
Mr. Gregory Watt* melted seven hundred weight
of basalt, and kept it in the furnace several days
after the fire was reduced. It fused into a dark-
coloured vitreous mass, with less heat than was
necessary to melt pig-iron; as refrigeration proceeded, the mass changed into a stony substance, and
globules appeared; these enlarged till they pressed
laterally against each other, and became converted
into *polygonal prisms*. The articulated structure
and regular forms of basaltic columns have, there-

* Philos. Trans. 1804.

fore, resulted from the crystalline arrangement of the particles in cooling; and the concavities, or sockets, have been formed by one set of prisms pressing upon others, and occasioning the upper spheres to sink into those beneath; thus the different layers of spheres have been articulated together, as in these specimens of basaltic columns from the Giants' Causeway.

1   2   3

TAB. 75.—BASALTIC COLUMNS, FROM THE GIANTS' CAUSEWAY.

Fig. 1. A block partially decomposed, exhibiting the primitive spheroidal figure of the prism. 2. Portions of columns. 3. The concave surface of a joint.

Proofs of the correctness of this inference are afforded by the occurrence of basaltic fragments, in which a sphere is enveloped by a polyhedral figure; and from the fact, that when basalt is not divided into regular prismatic columns, it often forms laminated spheroids, which, varying in size, constitute by aggregation extensive masses of rock.

T T

Before passing to the examination of the unstratified rocks, I will digress for a few moments, to
direct your attention to some highly interesting
examples of basaltic rocks, and the changes effected
on the contiguous strata, by the intrusion and contact of these ancient lava currents.

TAB. 76.—STAFFA.

(Drawn by Miss Duppa.)

24. STAFFA—FINGAL'S CAVE.—Many of the
Hebrides (or Western Isles of Scotland) are
almost wholly composed of trap rocks. The island
of Staffa* is the most celebrated, from a chasm or
recess in the rock, which has been produced by the
degradation and removal of the basaltic columns by
the waves. This natural cavern is of singular
beauty, and is well known by the English name of
Fingal's Cave, but it is called by the islanders

* Staffa, a Norse term, signifying staff or column.

Naimh-bim, or cave of music, from the murmuring
echoes occasioned by the surges, which, in rough
weather, dash with violence into the chasm. To the
elegant and effective pencil of Miss Duppa I am
indebted for the beautiful paintings of this singular
cave, and of the island (Tab. 76), with which these
remarks are illustrated.

Staffa is a complete mass of basalt, covered by a
thin layer of soil; it is about two miles in circumference, and is surrounded on every side by steep
cliffs, about seventy feet high, formed of clusters of
angular columns, possessing from three to six or
seven sides. Fingal's Cave, first made known to the
public by Sir Joseph Banks, in 1772, is on the
south-east corner of the island, and presents a magnificent chasm 42 feet wide and 227 in length.
The roof, which is 100 feet high at the entrance,
gradually diminishes to 50, and is composed of
the projecting extremities of basaltic columns;
the sides, of perpendicular pillars; and the base,
of a causeway of the same materials. The vaulted
arch presents a singularly rich and varied effect; in
some places, it is composed of the ends of portions
of basaltic pillars, resembling a marble pavement;
in others, of the rough surface of the naked rock;
while in many, stalactites mingle with the pillars in
the recesses, and add, by the contrast of their colours, to the pictorial effect, which is still farther
heightened by the ever varying reflected light thrown
from the surface of the water, which fills the bottom

of the cave. The depth of the water is nine feet, and a boat can therefore reach the extremity of the cave in tolerably calm weather; but when the boisterous gales of that northern clime drive into the cavern, the agitated waves dashing and breaking among the rocky sides, and their roar echoed with increased power from the roof, present to the eye and ear such a scene of grandeur as bids defiance to any description. The short columns composing the natural causeway before mentioned, continue within the cave on each side, and form a broken and irregular path, which allows a skilful and fearless climber to reach the extremity on the eastern side on foot: but it is a task of danger at all times, and impossible at high tide, or in rough weather. It would be useless, observes Dr. MacCulloch, to attempt a description of the picturesque effect of a scene which the pencil itself is inadequate to portray. But even if this cave were destitute of that order and symmetry, that richness arising from multiplicity of parts, combined with greatness of dimension, and simplicity of style, which it possesses; still the prolonged length, the twilight gloom half concealing the playful and varying effects of reflected light, the echo of the measured surge as it rises and falls, the transparent green of the water, and the profound and fairy solitude of the whole scene, could not fail strongly to impress a mind gifted with any sense of beauty in art or in nature.*

* MacCulloch's Western Isles.

The basalt, of which the columns are composed, is of a dark greenish-black hue, highly coloured by iron ; a thin layer of silicious cement is seen between the joints, or articulations, which is called mortar by the islanders, and strengthens their persuasion that this wonderful cave is the work of art.

25. THE GIANTS' CAUSEWAY. — In the sister kingdom, a magnificent range of basaltic pillars appears on the northern coast of Antrim. It consists of an irregular group of hundreds of thousands of pentagonal, jointed, basaltic columns, varying from one to five feet in thickness, and from twenty to two hundred feet in height. The structure of these masses I have already described ; their prevailing colour is a dark greenish-grey. In the cliffs, a chasm, formed by the inroads of the waves, presents a natural cavern, about sixty feet high, and of great picturesque effect; the entrance is nearly thirty feet in width, and the walls are formed of dark basalt. But the great interest of this spot, in a geological point of view, is the altered structure observable in the sedimentary rocks wherever they have been traversed by the basalt.

26. ROCKS ALTERED BY CONTACT WITH BA-SALT.—I have frequently had occasion to allude to the changes effected in sedimentary strata by the intrusion of basalt, and other volcanic rocks ;* but I have reserved for the present occasion a more particular exposition of the phenomenon. The chalk in the

* See page 256.

north of Ireland constitutes a line of cliffs traversed by basalt, which sometimes forms vertical dykes, and at others extensive beds, having a columnar structure. The chalk is about 270 feet thick, and rests on a green sandstone, called *mullattoe*, the equivalent of the *glauconite*, or firestone (page 273); it contains flint nodules, ammonites, belemnites, echinites, terebratulæ, and the usual fossils of the cretaceous formation. In the Isle of Rathlin, nearly vertical dykes of basalt are seen intersecting the chalk (as in this sketch, Pl. 4, fig. vi.*), which at the line of contact, and to an extent of several feet from the wall of the dyke, is completely changed. Those portions of the chalk which have been exposed to the extreme influence of the lava, are now a dark-brown crystalline rock, the crystals running in flakes, like those of coarse primitive limestone; the next state is saccharine — then fine-grained and arenaceous; a compact variety, with a porcellaneous aspect, and of a bluish-grey colour, succeeds; this gradually becomes of a yellowish-white, and passes insensibly into unaltered chalk. The flints in the indurated chalk are of a yellowish, or deep-red colour; the chalk is highly phosphorescent. The fossils are much indurated, but retain their usual structure.†

* Geological Transactions, vol. iii.

† My collection contains a suite of specimens illustrative of these various states, presented by G. B. Greenough, Esq., and Mr. Bryce, of Belfast.

To the south of Fair Head, in the county of Antrim, syenite (page 615) traverses mica schist and chalk ; and fragments of the latter are found broken up, and impacted in the erupted mass ; the included portions being changed into marble. The geological relations of that part of Ireland are as follow : 1, mica slate ; 2, coal shale, and new red sandstone ; 3, chalk.*

At Straithaird, in the Isle of Sky, vertical dykes and veins of trap intersect the horizontal strata of sandstone, as is shown in this sketch (Pl. 4, fig. VII.) ; porphyry, and other ancient lavas, also occur in the same island, sometimes protruding through, and at others spread over, clay-slate, red sandstone, and shelly limestone.†

In some of the slate districts, where the trap has burst through and overflowed the strata, fragments of slate are found imbedded in the basalt, appearing to have been detached from the rock at the intrusion of the lava, and become enveloped while the latter was in a state of fusion.

27. GRANITE VEINS — ROCKS ALTERED BY GRANITE.—From this subject, I pass to the consideration of the changes produced by granite and other ancient mineral masses that have been erupted in a melted state. The transition from granite to modern porphyritic trachytes, passes through infinite gradations, but all the modifications appear referrible to the degree of incandescence of the materials,

* Mr. Griffiths.　　　† Dr. MacCulloch.

the circumstances under which they were erupted, and their slow or rapid refrigeration. An instructive example of the passage of granite into basalt, described by Dr. Hibbert, will illustrate these remarks. In one of the Shetland Isles, a bed of basalt, extending for many miles, is seen in contact with granite. At a little distance from the junction of the rocks, the basalt contains minute particles of quartz, and these become larger and more distinct as they approach the granite; hornblende, felspar, and greenstone (the latter is a homogeneous admixture of hornblende and felspar) next appear; still nearer, the rock consists of felspar, quartz, and hornblende: and at the line of junction, felspar and quartz form a mass, which requires but the presence of mica to be identical with the granite in which it is insensibly lost.*

Veins, I have previously stated (page 177), are fissures or chasms, originating either in mechanical disturbance, or from contraction in mineral masses during their consolidation; and the mechanical veins are found filled by subsequent infiltrations or depositions. It is obvious that these veins must be of later origin than the rock which they traverse. Thus the veins of granite, represented in this diagram, (Pl. 3, fig. 20,) are more modern than the mass through which they are disseminated. It therefore follows that when rocks of granite are intersected by veins, or dykes of the same substance, the latter

* Edinburgh Journal of Science.

are of later origin than the former, and have been injected into rents and openings of pre-existing granite
rocks; a proof that the formation of granite has taken
place at more than one epoch.   By numerous observations of phenomena of a like nature, it is now
clearly established that melted granite has been
ejected during the Cumbrian, Silurian, carboniferous,
oolitic, chalk (see Pl. 4, fig. VII.), and even tertiary
epochs. When granite appears to have been erupted
among the secondary strata, the latter, as we have
already remarked, are invariably altered near the
line of junction; but when consolidated masses of
granite have been protruded, no such change is
observable.   Into the slate rocks of the Cumbrian
chain, syenite, porphyry, and greenstone, have been
injected in a melted state, and now fill up the fissures
produced during the general movements of those
strata; but the central nucleus of primary rock exhibits no such appearance.   In Cornwall, and other
places, the granitic rocks were evidently in a state of
fusion, for the slates are penetrated by veins of granite;
and in some instances are changed into fine-grained
mica, or hornblende slate.   An extraordinary fact is
noticed by M. Elie de Beaumont.   In the environs
of Champoleon, where granite comes in contact
with the Jura limestone, whatever may be the
position of the surfaces in contact, the limestone
and the granite both become metalliferous near the
line of junction, and contain small veins of galena,
blende, iron and copper pyrites, &c.; and at the

same time the secondary rocks are harder and more crystalline, while the granite has undergone a contrary change.*

28. METAMORPHOSED ROCKS. — Enough has been advanced to convey a general idea of the character and relation of the primary crystalline rocks, and of the agency which has reduced them to their present state; but the question naturally arises —What was their original nature? Intense heat has effected the present arrangement of their molecules, but upon what materials was that influence exerted? The transmutation, by heat, of chalk into crystalline marble—of loose sand into compact sandstone—of argillaceous slate into porcelain jasper—of coal into anthracite—of anthracite into shale and slate—of slate into micaceous schist—of micaceous schist into gneiss and granite—of the latter into trap—and so forth—together with the characters presented by the mineral products of existing volcanoes, prepare the mind to receive without surprise the theory of an eminent geologist and chemist, M. Fournet, *that all the primary rocks are simply sedimentary deposites metamorphosed by igneous action.*† I will only add that this opinion is but a modification of that long since expressed by our illustrious countryman, Hutton, that granite rocks

* De la Beche.

† The general reader will find an interesting account of M. Fournet's theory in Jameson's Edinburgh Journal, No. xlvii. p. 3.

are consolidated and altered sediments which have accumulated at the bottom of the ocean.

29. METALLIFEROUS VEINS.—In my description of the fissures observable in consolidated strata, I mentioned that the great depositaries of the metals are found in certain cavities termed metalliferous veins; which are separations in the continuity of rocks, of a determinate width, but extending indefinitely in length and depth, and more or less filled with metallic and mineral substances of a different nature from the masses they traverse. These natural stores of hidden treasures are not confined to any epoch of formation, nor to any tracts of country, although most frequent in beds that form mountain elevations, and in the oldest rocks. I have already mentioned, (page 258,) that veins of iron, copper, arsenic, silver, and gold, occur in tertiary strata. Many veins are fissures of mechanical origin, into which metalliferous matter has been sublimed from the effects of high temperature; but others have resulted from an electro-chemical "separation, or segregation, of certain mineral and metallic particles from the mass of enveloping rock, while it was in a soft or fluid state, and their determination to particular centres." The nature of these veins receives illustration from the nests of spar and mineral matter in masses of trap rocks from Scotland, in which, as you perceive, there was no possibility of the introduction of any foreign substance from without. From the observations

of M. Fournet, in the mines of Auvergne, it seems
probable that sulphurets of iron, copper, lead, zinc,
barytes, and other minerals, have been introduced
at different periods, by electro-chemical action, accompanied by new fractures and dislocations of the
rocks, and the widening of previous fissures.*

There appear to be certain associations of metallic
substances in the veins; as for instance, iron and
copper, lead and zinc, tin and copper;† and those
ores which are combined with a similar base, as
sulphurets, carbonates, phosphates, arseniates, &c.
are commonly found together.‡   The following is
a brief notice of the geological distribution of a few
of the chief metals.

Lead.—The ores of this metal are very numerous : and the sulphuret, or galena, occurs in primary and secondary rocks.

Tin—exists in veins traversing granite and schist ; those of
Cornwall have been celebrated from the earliest historical
period.

Copper—is found in primary and secondary rocks, and is often
native, that is, in a pure metallic state ; and crystallized.

Gold—exists in granite and quartz rocks.   The gold found in
the mud and sands of rivers (as these grains from Ovoca in
Ireland, collected by the late Earl of Chichester,) is derived
from disintegrated rocks.

Silver.—This metal is found in transition and primary rocks ;
sometimes native (as in these specimens from Cornwall, from
the mines of my friend, John Hawkins, Esq. of Bignor Park);
and in ores associated with arsenic, cobalt, &c.

The almost universal presence of the ores of iron,
and the infinite variety of its combinations, are well

* Mr. Lyell's Anniversary Address.
† Mr. Burr.   ‡ Professor Phillips.

known. The formation of what is termed bog-iron ore, found in marshes and peat-bogs, is supposed to have been derived from the decomposition of rocks over which water has flowed; but the observations of Ehrenberg, to which I shall presently advert, seem to indicate a different origin.

30. COPPER ORE OF NEW BRUNSWICK.*—An illustration of a metallic deposit by the effects of chemical action, without the agency of heat, is afforded by a singular formation of copper ore, which occurs in New Brunswick. In a bed of lignite, which is covered by a few feet of alluvial soil, and rests on a conglomerate, the precise nature of which is not stated, occurs a nearly horizontal layer of green carbonate of copper, about eight inches in thickness. The ore is disseminated through the lignite in the same manner as the metallic ores are usually blended with their accompanying vein-stones. This bed bears a close analogy to the modern cupreous deposits of Anglesea, or of some parts of Hungary and Spain, where, at the present time, water charged with copper in solution, is by the introduction of iron made to precipitate the former metal. From the stratum of lignite occurring with the copper, and the mode in which the latter is interspersed throughout the mass, it would appear that the water in which the vegetable matter floated was, at the same time, saturated with a solution of

* Mining Review, vol. iv. No. 4. By Frederick Burr, Esq.

copper, and that both the organic and mineral substances subsided to the bottom together, and formed
the singular compound deposit under consideration,
over which, probably at a subsequent period, the
alluvial covering was drifted.

31. SAPPHIRE, RUBY, EMERALD. — Connected
with the changes to which the metamorphic rocks
have been subjected is the formation of some of
those minerals, which, from their beauty, splendour,
and use as ornaments, are termed precious-stones.
The sapphire and oriental ruby, which are prized
next to the diamond, and almost equal that gem in
hardness, are found in trap rocks; and the common
corundum, which is a species of the same mineral,
and the emerald, occur in granite. The two former
principally consist of aluminous earth;* and the
supposition that they have been formed by intense igneous action, is not only probable, but is
rendered almost certain, by the late experiments of
M. Gaudin, who has succeeded in producing fictitious rubies, which, in every respect, resemble the
natural gems. These were formed by submitting
aluminum, with a small quantity of calcined chro-
mate of potash, to the influence of a powerful oxy-
hydrogen blowpipe, by which the materials were
melted into a crystalline mass, that presented, when
cooled, all the characteristics of the ruby.

* The sapphire affords, by analysis, 98·5 of alumine, 0·5
of lime, and 1 of oxide of iron; the ruby, 90 of alumine, 7 of
silex, and 1·2 of oxide of iron.—*Phillips's Mineralogy.*

I will only add, that instances occur in which garnets and other crystals are found in shale, when altered by contact with a dyke of igneous rock, though altogether wanting in every other part of the bed; a proof that they have been produced by the effects of heat on those parts of the sedimentary deposits which were most exposed to the influence of the erupted mass.*

32. REVIEW OF THE SILURIAN AND SLATE SYSTEMS.—Let us now review the leading phenomena which have been brought under our notice in this discourse. The Silurian system presented all the usual characters of sedimentary deposites, with which our previous investigations have rendered us familiar. Its marine origin is evinced by the organic remains; and the strata have evidently been formed and consolidated by mechanical, chemical, and vital agency, acting through a long period of time, in like manner as in the production of the newer secondary formations. The fossils consist of a few algæ, equiseta, lycopodiaceæ, and ferns; about ninety species of polyparia, of which the lamellar and cellular corals form by far the largest proportions; thirty-five species of crinoidea; about two hundred and sixty species of bivalve shells, and eighty of cephalopoda; sixty-five species of trilobites, or other crustacea; and the remains of a few species of fishes. Mr. Murchison, from his extensive collection of fossils, has selected several which he considers

* Mr. Lyell.

characteristic of the four groups into which he has subdivided the system; and these fossils, with but few exceptions, are assumed to be specifically distinct from those of the carboniferous group.

In the Cumbrian, or slate system, we have a vast argillaceous formation, with numerous conglomerates; and from the structure of the entire series, it would appear that after the deposition of the strata by water, the whole had been exposed to the long-continued influence of heat, by which the original sedimentary character was either greatly modified, or entirely obliterated; about twenty or thirty species of shells and corals, consisting of cyathophyllia, spiriferæ, productæ, &c. are the only organic remains. In accordance with the slaty structure, is the prevalence of melted rocks throughout the Cumbrian epoch; for not only do granite, porphyry, serpentine, and trap, occur in veins and dykes, but also intercalated with the strata, as if the melted matter had been poured over argillaceous sediments at the bottom of the sea, and had become covered by succeeding deposites.

We have thus, in these two systems, evidence of marine depositions going on through an immense period of time, during which the sea abounded in polyparia, mollusca, and crustacea; for although organic remains prevail only in the uppermost or newest group, yet as we have decided proof that the lowermost division has been subjected to intense heat, and that even the lines of stratification

are in a great measure melted away, it is clearly
reasonable to conclude, that the absence of fossils
is attributable to the obliteration of the remains of
the animals which lived and died in the waters
that deposited the slate. We must not, however,
fail to remark, that the relics of organized beings
which remain are of·a peculiar type, and altogether
different from the corals and shells of the newer
secondary formations.

33. REVIEW OF THE METAMORPHIC ROCKS.—
The traces of stratification, a structure which, we
have seen, is characteristic of aqueous formations
(page 174), are evident in the upper group of the
crystalline metamorphic rocks ; and there is also an
obscure resemblance to the alternate depositions of
secondary beds, in the succession of different mineral masses, as gneiss, mica schist, quartz rock, &c.
But in the lowermost term of the series, the granite,
even these apparent relations to the stratified formations, are wanting ; and in the amorphous masses,
veins, and dykes, we have the effect of long continued and intense igneous action, produced under
circumstances which have given to the resulting
rocks a very peculiar character. There is one
striking deduction which M. Fournet has drawn
from the mineralogical character of these rocks,
namely, that those masses which, according to our
chemical knowledge, would require the most intense
and long continued incandescence for their formation,
—those in which quartz largely predominates,—are

precisely those which from their geological position must have been longest exposed to such an
agency—hence, in granite, the foundation rock,
quartz, which is the most infusible and refractory
material, largely prevails. The possibility of an
earth being converted by intense heat into the
hardest and purest crystal, was shown in the formation of fictitious rubies. To the granite succeed
rocks in the exact order of their containing less
quartz, and being therefore more easily fusible—
as granite with a large proportion of felspar, porphyry, serpentine, mica schist, and clay slate.* If
we take these phenomena into consideration, toge-
gether with the facts previously stated, of the transmutation of one substance into another by heat, it
appears to me, that in the present state of our knowledge, we are warranted in concluding that granite
and its associated rocks, are nothing more than
sedimentary deposites altered by igneous agency.

34. ORGANIC REMAINS IN THE METAMORPHIC
ROCKS??—I have before stated, that with the last
of the slate rocks, all traces of organization are
lost; but this assertion requires some modification.
Let us now resume the inquiry. From the intense
heat to which the metamorphic rocks have been
exposed, we cannot expect to find any elementary
organic structures, except those which are formed
of materials capable of resisting the effects of such
an influence. The observations and experiments of

* Jameson's Edinburgh Journal, No. 47.

Mr. Reade have shown (page 565), that vegetables possess a structure which is composed of silex, and is indestructible in a common fire. In animals, we seek in vain for an elementary tissue, capable of resisting the powerful influence of heat, except in those minute beings, the· infusoria, of which I treated in a previous lecture (page 300). In certain families of these living atoms, the soft body of the animal is protected by two cases or shields, like the cypris (page 346) ; and these cases are, in various species, composed of lime, iron, or flint. In others, the skeleton or solid support is in the form of rings, or moniliform (bead-like) threads. The silicious skeletons and shields of the infusoria, are therefore the only animal structures that can escape destruction, in substances subjected to the influence of a high temperature ; and it is clear, that if the skeletons or durable parts of any other animals were exposed to such an agency, all traces of their organization would be obliterated. It would therefore be a hopeless task to seek for any trace of animals, in rocks where even the lines of stratification are melted away, except of those which, like the infusoria, possessed silicious skeletons. When speaking of the fossils of the chalk, it was stated that the coatings of many of the flints contained myriads of the silicious skeletons of animalcules, and that some rocks were almost wholly composed of such remains (page 300). Mr. Reade has recently favoured me with some remarks on this subject illustrated by drawings of organic bodies found in flint.*

Ehrenberg, to whom we are largely indebted for opening this new field of inquiry, has discovered the remains of this class of animals in numerous deposites. Thus the ferruginous or ochreous film or scum seen on the water of marshes, or of stagnant pools, or collected at the bottom of ditches, sometimes forming a red or yellowish mass many inches thick, without any consistence, which divides upon the bare touch into minute atoms, and when dried, resembles oxide of iron, is found to be wholly composed of the shields of infusoria *(gaillonella ferruginea)*. The formation of bog iron-ore is supposed to be in a great measure dependent on these animals. A ferruginious mass from a peat bog, "which appears to have owed its origin to the action of volcanic heat at the bottom of the sea," entirely consisted of shields of infusoria *(naviculæ)*. The semi-opal, and the tripoli of the tertiary deposites, are wholly composed of the fossil remains of this class of animals. In the secondary formations, we have seen that they are equally abundant. Ehrenberg also distinctly states, that while in the instances above mentioned, there cannot be the least doubt of the nature of the organic remains; in the semi-opal of the serpentine formation of Champigny, and in the precious opal of the porphyry, he has detected bodies so exactly

* Appendix L.

similar, that although at present he hesitates posi-
tively to affirm that they are organic, he can scarcely
entertain any doubt upon the subject.   I will now
place before you Mr. Reade's remarks, and his
drawings of the apparently organic bodies in mica
schist.*   At present, my information on this highly
interesting inquiry extends no farther.

35. RELATIVE AGE OF MOUNTAINS.—We have
seen that the intrusions of melted rocks have not
only altered the chemical nature of the strata,
through which they were erupted, but have also
changed their position and relations, and produced
corresponding modifications in the physical geography of the dry land, transforming plains into
mountain peaks, and occasioning the subsidence of
elevated regions to the bottom of the deep.   As
these revolutions took place at various epochs,
separated from each other by considerable, or brief
periods of repose, it is manifest that the existing
mountain chains are of very different ages.   By a
careful examination of the phenomena which bear
upon this question, the relative antiquity of many
of the principal ranges has been determined ; or,
in other terms, it has been ascertained during what
geological epochs the Alps, Pyrenees, Andes, &c.

* These are inserted in the Appendix L.   I would refer the
reader, whose curiosity is awakened by these remarks, to the
Third Part of Mr. Taylor's Scientific Memoirs, for a translation
of two of Ehrenberg's Memoirs on Fossil Infusoria.   London,
1837, price 6s.

were elevated above the waters. My observations on this subject must, however, be restricted to an explanation of the mode of induction employed, and a brief notice of some of the results which have been obtained. The positions and relation of the secondary strata afford the principal data by which this problem may be solved ; for, as secondary and tertiary formations have been deposited in directions either nearly or entirely horizontal, it is obvious, that when they are found highly inclined, and in contact with mountain masses of primary or volcanic rocks, the latter must have been protruded *since* the sedimentary were deposited, and of course during the secondary or tertiary epochs, as the case may be. On the contrary, if we find other strata in contact with the same masses, but only touching them with their edges, or encircling their base, it is obvious that the mountains must have been elevated before the formation of the surrounding deposites. It is by cautious induction of this kind, that a distinguished savant, M. Elie de Beaumont, has shown—1. That the mountains of Erzegebirge, in Saxony, and of the Côte d'Or in Burgundy, are newer than the Jura limestone, but older than the green sand and chalk. 2. That the Pyrenees and Appennines are of about the same age with the chalk formation. 3. That the western part of the Alps is newer than the older tertiary formations, and was raised up after the last of the newer pliocene strata were deposited. It is obvious that the protrusion of such immense masses as the Alps or Pyrenees from the bottom of the sea, must have dislodged a vast body of water, and created a series of waves high and powerful enough to cause transitory but destructive inundations over such portions of the adjacent dry land, as ·were only a few hundred feet above the level of the sea.   In this way M. Beaumont thinks some of those revolutions may have been produced, which seem to have extinguished animal life at different periods, and prepared the way for new forms of living beings.

36. SUCCESSION OF CHANGES IN THE ORGANIC KINGDOMS.—I now approach the termination of this argument, and it will be instructive to review the phenomena which have passed before us, in order that we may fully comprehend and retain a clear conception of the leading principles and inferences of geology.   To condense my remarks as much as possible, I propose offering—*firstly*, a summary of the changes which have taken place in the animal and vegetable kingdoms, and, *secondly*, in the varying physical conditions of the earth's surface, during the vast periods which our investigations have embraced ; and, *thirdly*, to consider the influence exerted by vital action in the elaboration of the solid materials of the globe.   I shall conclude by a few general observations on the highly important and deeply interesting facts, that have formed the subject of our contemplations.

So numerous and varied have been the phenomena

examined, that with the view to recall the principal
events which have marked our progress, I place
before you the series of illustrations employed in
these lectures, that you may perceive at a glance the
striking contrast presented by the fauna and flora
of different geological epochs.* In the first stage,
traces of the existing orders of animated nature
were everywhere apparent; and works of art, with
the bones of man, and the remains of vegetables
and of animals, were found in the modern deposites. In the succeeding era, many species and
genera, both of plants and animals, were absent.
Large terrestrial pachydermata greatly predominated, and the vegetation was principally of a character referrible to temperate and intertropical
climes ; while the seas abounded in fishes, crustacea, and mollusca, as at the present time.

The next epoch presented one wide waste of
waters, teeming with the general types of marine
beings, but of different species and genera to those
of the previous eras, and bearing a large proportion
of cephalopodous mollusca. A few algæ and fuci
made up the marine flora; and drifted trunks of
coniferæ and dicotyledonous trees, and a few reptiles,
were the only indications of the dry land and its

* The reader may realize this idea by referring to the illustrations of these volumes, commencing with the fossil human
skeleton (p. 63), and proceeding from the large mammalia
(pp. 127, 133, 148), to the last of the series, the corals and
shells of the ancient secondary deposites (pp. 421, 501).

inhabitants. The delta of a mighty river now appeared, containing the spoils of an extensive island or continent; and the remains of colossal reptiles, and of unknown forms of tropical plants, marked the era of the country of the Iguanodon.

We were then conducted to other seas, whose waters abounded in fishes and mollusca, and were inhabited by marine reptiles, wholly unlike any that now exist; while the dry land was tenanted by enormous terrestrial and flying reptiles, marsupial animals and insects, and possessed a tropical flora of a peculiar character. In the next era, we found another sea swarming with fishes, mollusca, and corals, and with reptiles similar to those of the preceding period.

The succeeding change disclosed extensive regions, covered by a luxuriant vegetation; with groves and forests of palms, arborescent ferns, and coniferæ, and gigantic trees related to the existing club mosses and equisetaceæ; the numerical preponderance of the flowerless plants, constituting a character wholly unknown in modern floras. The ocean abounded in mollusca, radiaria, and crustacea, of genera and species unlike any that had previously appeared.

We advanced to other oceans, swarming with polyparia, mollusca, radiaria, and fishes, which bore a general analogy to those of the preceding seas, but belonged to different species: interspersions of cryptogamous plants, with a flora related to the one immediately antecedent, marked the existence of dry land. But traces of animal and vegetable existence became less and less manifest, and were at

666 WONDERS OF GEOLOGY.

length reduced to a few shells, corals, and sea-weeds; these finally disappeared, and dubious indications of infusoria were the last vestiges of organic life.

37. SUCCESSIVE DEVELOPMENT OF THE ORGANIC KINGDOMS.—If we reverse the order of the argument, and pass in succession from the ancient to the modern epoch—from the regions of sterility and desolation, to those in which animal and vegetable life were profusely developed—we obtain the following results:—

Geological Formations.	Character of the Fossil Fauna.	Character of the Fossil Flora.
Granite	*Infusoria ??*	No traces of vegetables
Lower slate syst.	*Corals* and *shells (brachiopoda)*	Fuci ?
Upper slate syst.	*Corals, crinoidea,* shells, and *trilobites*..................	Fuci.
Silurian system .	Corals, crinoidea, *orthocera,* and other shells, trilobites, *fishes*................	Fuci.
Carboniferous system	Corals, crinoidea; cephalopoda, shells, both marine (chiefly brachiopoda) and freshwater; trilobites, *insects, sauroid fishes, reptiles*........	Several hundred species of plants; the *vascular cryptogamia* largely developed. Palms, tree-ferns, coniferæ. Dicotyledonous plants very rare.
Upper secondary.	Corals and shells of all orders; crinoidea, fishes, insects, belemnites, ammonites, &c. *Reptiles,* both marine and terrestrial, of numerous genera and species; and many of gigantic size. One genus of *marsupial mammalia—Didelphis;* and one of *birds—Ardea*............	*Zamiæ, Liliaceæ.* Palms. Tree ferns. Coniferæ. Dicotyledonous trees rare.
Tertiary	*Terrestrial herbivorous,* and *carnivorous mammalia.* The numerical proportion of reptiles comparatively small. Birds, fishes, and all the existing orders............	Dicotyledonous trees prevail; coniferæ; palms, tree ferns, &c.
Modern epoch ...	MAN, and contemporary animals	Remains of the existing vegetation.

This rapid sketch presents but an outline of the most striking changes observable in the organic forms preserved in the several formations of the sedimentary deposits. In this view—setting apart the infusoria—a few fuci, mollusca, and polyparia are the first evidence of creation; these are followed by a larger development of the same orders, and the addition of crinoidea, crustacea, and fishes; in the succeeding period reptiles and insects appear, with sauroid fishes, and an immense development of vegetable life, particularly of the cryptogamic class. Large reptiles next prevail to an extraordinary degree; and one genus of birds, and one of mammalia, attest the existence of the higher orders of animals. The vegetable kingdom is greatly modified; and plants related to the zamiæ and to the liliaceæ preponderate, with coniferæ and dicotyledonous trees. The next remarkable change is in the sudden increase of mammiferous animals, and the reduction of the reptile tribes; the large pachydermata, as the mammoth, elephant, &c. first appear. From this period till the creation of man, there are no striking general modifications in the various orders of animal and vegetable existence.

It was from this *apparent* successive development of living beings, from the most simple to the most complex organizations, that the geological theory of periodical creations which once prevailed, took its rise; but I scarcely need remark, that the facts we have stated warrant no such inference: we have

seen that the minutest living atom possesses a
structure as wonderful as our own; and that some
of the fossil animals which first appear in the strata,
belong to families with a highly developed organization.   Nor does the vegetation of the earliest
periods lend any support to such a hypothesis.
Fungi, lichens, hepatica, or mosses, as I have before
remarked, do not form the flora of the carboniferous
strata, but coniferæ, and the most perfectly organized
of the cryptogamic class.

38. GEOLOGICAL EFFECTS OF MECHANICAL
AND CHEMICAL ACTION.—The physical changes
that have taken place in the earth's surface, are in
perfect harmony with the modifications observable
in animated nature; for the laws of mechanical and
chemical action are indissolubly connected with
those which govern vital phenomena; and we have
incontrovertible evidence, that throughout the vast
periods over which geological speculations extend,
the same causes have operated, the same effects
followed.   Thus, heat and cold, drought and moisture, and other atmospheric influences, have dissolved the loftiest peaks—rivulets and torrents have
eroded the sides of the mountain-chains—streams
and rivers have worn away the plains, and carried
the spoils of the land into the bed of the ocean—
the waves of the sea have wasted its shores, and
destroyed the cliffs and rocks which opposed their
progress—silt has been changed into clay—calcareous mud into limestone—sand into sandstone—

pebbles into conglomerates and breccia—and animal and vegetable remains have been imbedded, and added to the mineral accumulations of countless ages.

Beneath the surface, the action of electro-chemical forces has been alike unintermitting—vegetable matter has been converted into bitumen, coal, amber, and the diamond—earth into crystals—limestones into marble—clay into slate, and sedimentary into crystalline masses; the volcano has poured forth its rivers of molten rock—the earthquake rent the solid crust of the globe—beds of seas have been elevated into mountains—subsidences of the land and irruptions of the ocean have taken place— and the destructive and conservative influences of both fire and water, have been constantly exerted; the phases of action have alone differed in duration and intensity.

39. ROCKS COMPOSED OF ORGANIC REMAINS. —In a previous discourse I dwelt upon the highly interesting subject of the elaboration of solid materials from gaseous and fluid elements by vital action, and the formation of islands and continents by countless myriads of living instruments. It is my present purpose to consider how far the present solid materials of the earth's surface have been derived from organized beings. The processes by which animal and vegetable structures are converted into stone, and the various states in which their fossil remains occur, have already been explained.

The strata of vegetable origin consist of peat—
forests ingulfed by subsidences of the land, or imbedded in the mud of rivers and deltas, or in the
basin of the sea—the lignite and brown coal of the
tertiary deposites—the coal and shales of the carboniferous strata—and of the silicified and calcareous
trunks of trees in tertiary and secondary formations.
But the deposites which are derived either wholly or
in part from animal exuviæ, are so numerous, and
of such prodigious extent, that the interrogation of
the poet may be repeated by the philosopher—

"Where is the dust that has not been alive?"—YOUNG.

Probably there is not an atom of the solid
materials of the globe which has not passed
through the complex and wonderful laboratory of
life. Thus we find that all the orders of animals,
from the infusoria up to man, have more or less
contributed, by their organic remains, to swell
the amount of solid matter of the crust of the
earth. The following tabular arrangement presents
in a condensed form some of the most striking
results.

ROCKS COMPOSED WHOLLY OR IN PART OF
ANIMAL REMAINS.

Strata.	Prevailing Remains.	Formations.
Trilobite-schist	Trilobites	{ Silurian { system
Dudley limestone	Corals, crinoidea, trilobites and shells	—
Shelly limestone	Productæ, spiriferæ, &c.	—
Mountain limestone ...	Corals and shells	⎧Carboni- ⎨ ferous ⎩ system

Strata.	Prevailing Remains.	Formations,
Encrinital marble	Lily-shaped animals and shells	Carboni-ferous system.
Muscle-band..............	Fresh-water muscles	—
Ironstone nodules	Trilobites, insects, and shells	—
Lias-shales and clays ...	Pentacrinites, reptiles, fishes	Lias
Limestone..................	Terebratulæ, and other shells	—
Lias conglomerates......	Fishes, shells, corals	—
Gryphite limestone	Shells, principally gryphites............	Lias
Limestone..................	Terebratulæ, and other shells	Inferior oolite
Stonesfield slate	Shells, reptiles, fishes, insects	Oolite
Pappenheim schist	Crustacea, reptiles, fish, insects	—
Bath-stone..................	Shells, corals, crinoidea, reptiles, fishes	—
Limestone	Cephalopoda, principally ammonites	—
Coral-rag	Corals, shells, echini, ammonites ...	—
Bradford limestone	Crinoidea, shells, corals, cephalopoda	—
Portland oolite	Ammonites, trigoniæ, and other shells	—
Purbeck and Sussex marble	Fresh-water shells, crustacea, reptiles, fishes	Wealden
Wealden limestone	Cyclades, and other fresh-water shells, crustacea, reptiles, fishes ...	—
Tilgate grit (some beds)	Reptiles, fishes, fresh-water shells ...	—
Faringdon gravel	Sponges, corals, echini, and shells ...	Shanklin sand
Jasper and chert	Shells	—
Greensand..................	Fibrous zoophytes	—
Chalk	Corals, radiaria, echini, shells, fishes.	Chalk
Maestricht limestone ...	Corals, shells, ammonites, belemnites, and other cephalopoda—reptiles ...	—
Hippurite limestone ...	Shells, principally hippurites	—
Hard chalk (some beds)	Echini and belemnites	—
Flints	Sponges, and other fibrous zoophytes Infusoria, and spines of zoophytes ... Echini, shells, corals, crinoidea	— — —
Limestone	Fresh-water shells	Tertiary
Nummulite rock	Nummulites	—
Septaria....................	Nautili, turritellæ, and other shells	—
Calcaire grossier	Shells and corals	—
Gypseous limestone ...	Mammalia, (palæotheria, &c.) birds, reptiles, fishes	—
Silicious limestone	Shells	—
Lacustrine marl	Cyprides, phryganeæ, fresh-water shells	—

Strata.	Prevailing Remains.	Formations.
Monte Bolca limestone	Fishes	Tertiary
Bone-breccia	Mammalia, and land-shells	—
Sub-Himmalaya sand- stone	Elephant, Mastodon, &c. reptiles ...	—
Tripoli	Infusoria	—
Semiopal	Infusoria ....,......................	—
Guadaloupe limestone	MAN, land-shells and corals	Human epoch
Bermuda limestone......	Corals, shells, serpulæ	—
Bermuda chalk	Comminuted corals, shells, &c.	—
Bog iron ochre	Infusoria	—

This list might be almost indefinitely extended, for I have omitted numerous strata, in which animal remains largely predominate; and in the tertiary and modern epochs, every order of animated nature is found to have contributed, more or less largely, to the sedimentary deposites; the bones of man, &c. first appearing in the most recent accumulations; and by the geological causes now in action, not only the remains of the existing orders of living beings, but also works of human art, are added day by day to the solid crust of the globe.

40. GENERAL INFERENCES. — Restricting ourselves within the bounds of legitimate induction, and forbearing to speculate on those points which rest on insufficient or questionable data, we may nevertheless venture to draw some general inferences as to the varying physical conditions of our planet, and of animal and vegetable life, through the countless ages contemplated by geology.

From the remotest period in the earth's physical history recognizable by man, to the present time, the mechanical and chemical laws, which govern inorganic matter, appear to have undergone no change. The wasting away of the solid rocks by water, and the subsequent deposition and consolidation of the detritus by heat; the subsidence of the dry land beneath the sea, and the elevation of the ocean bed into new islands and continents; the decomposition of animal and vegetable substances on the surface, and their conversion into stone or coal, under circumstances in which the gaseous principles were confined; the transmutation of mud and sand into rock, and of earthy minerals into crystals, — these physical changes have been going on through all time, under the influence of those fixed and immutable laws, established by Divine Providence for the maintenance and renovation of the material universe.

And although among the sentient beings which have from time to time inhabited the earth, we discover at successive periods the appearance of new forms, which flourished awhile and then passed away, while other modifications of life sprung up, and after the lapse of ages, in their turn were annihilated; yet the laws which governed their appearance and extinction, were in perfect harmony with those which regulate inorganic matter. Every creature was especially adapted to some peculiar state of the earth at the period of its development; and when the physical conditions were changed, and no longer favourable for the existence of such a type of organization, it necessarily became ex-

x x

tinct.* Thus we have seen different modifications of animal and vegetable life prevailing at different epochs of the earth's physical history, yet all presenting the same principles of structure, the same unity of purpose; all bearing the impress of the same Almighty hand. The creation of man, and the establishment of the existing order of things —which we are taught both by revelation and by natural records took place but a few thousand years ago,—are events beyond the speculations of philosophy.

It follows, from what has been advanced, that both animate and inanimate nature, linked together by indissoluble ties of mutual adaptation, have been governed by the same mechanical, chemical, and vital laws, from the earliest geological periods to the present time; and that the absence of the fossil remains of whole orders of animals in the remotest periods, although in some cases attributable to the feeble development of those types of being, may have been also dependent on the obliteration of their remains in the igneous rocks by high temperature: at the same time we must not forget, that we are examining ancient ocean beds, and may not yet have explored such parts of their vast abysses as may conceal the spoils of the land. I need not add, that the assumption of the successive evolutions of new forms of being,

* See page 103.

adapted to peculiar conditions, is only modified, not weakened, by this argument.

What, then, is the result of our inquiry into the ancient condition of our globe ?—That, so far as our present knowledge extends, all the changes produced by mechanical, chemical, or vital agency, whether on the surface or in the interior of the earth, have been taking place from the earliest geological periods; and, as like causes must produce like effects, will continue to take place so long as the present material system endures. Thus deposites now in progress may subside to the inner regions of the earth, and by exposure to long continued igneous action, all traces of sedimentary origin may be destroyed ; and at some distant period, the metamorphosed masses may appear on the surface in the form of peaks of granite, bearing with them the accumulated spoils of countless ages. I cannot, therefore, concur in the opinion of those, who imagine that in the granite we see the primeval solid framework of the earth—a consolidated crust which formed on the surface of a cooling planet, and was subsequently broken up by changes in the temperature of the earth. The only legitimate inference appears to be, that at a certain depth the solid materials of our globe become so entirely changed, as to afford no satisfactory data as to any antecedent period. In no department of natural inquiry is the admirable caution of the philosopher more necessary than in geology,—" that we should remember, knowledge is a temple, of which the vestibule only has been entered, and we know not what is contained within those hidden chambers, of which the experience of the past can afford us neither analogy nor clue."

41. FINAL CAUSES.—Geology, then, does not affect to disclose the first creation of animated nature; it does not venture to assume that we have evidence of a beginning; but it unfolds to us a succession of events, each so vast as to be beyond our finite comprehension, yet the last as evidently foreseen as the first. It instructs us, " that we are placed in the middle of a scheme—not a fixed, but progressive one—every way incomprehensible; incomprehensible in a measure equally with respect to what has been, what now is, and what shall be hereafter."*

The new page in the volume of natural religion, which Geology has supplied, has been so fully illustrated by Dr. Buckland, in his celebrated Essay, that I need not dwell at length on the evident and beautiful adaptation of the organization of numberless living forms, through the lapse of immense periods of time, to every varying physical condition of the earth, by which its surface was ultimately fitted for the abode of the human race— we have seen that the infusoria lived and died in countless myriads, and furnished the tripoli and the opal—that river-snails and sea-shells elaborated the marble for our temples and palaces, and polyparia

* Bishop Butler.

the limestone of which our edifices are constructed ; and that grass, herb, and tree, have been converted either into materials to enrich the soil, or a mineral which should serve as fuel in future ages when such a substance became indispensable to the necessities and luxuries of civilized man. Thus it is that geology has thrown a new interest around every grain of sand, and every blade of grass ; and that the pebble, rejected by the moralist and the divine,* becomes in the hands of the philosopher a striking proof of Infinite Wisdom.

But ought we to rest content in the assumption that all these wonderful manifestations of Creative Intelligence were solely designed to contribute to our physical necessities and gratifications?—Say, rather, that this display of beauty, power, and goodness, was designed to fill the soul with high and holy thoughts — to call forth the exercise of our reasoning powers—to excite in us those ardent and lofty aspirations after truth and knowledge, which elevate the mind above the sordid and petty concerns of life, and give us a foretaste of that high destiny, which we are instructed to hope may be our portion hereafter !

42. ASTRONOMICAL RELATION OF THE SOLAR SYSTEM.—Having thus endeavoured to interpret the natural monuments of the earth's physical history, let us contemplate the relation of our planet

* Paley. The remark alludes to the celebrated argument of this distinguished author.

to the countless orbs around us. For while astronomy explains that the solar system once existed
as a diffused nebulosity, which passing through
various states of condensation, formed a central luminary, and its attendant planets; it also
instructs us, that our system is but one incon-
siderable cluster of orbs, in regard to the group of
stars to which it belongs, and of which the milky-
way appears to be, as it were, a girdle; our system
being placed in the outer and less stellular part of
the zone.*   But the astounding fact, that all our
visible universe is but an aggregation, a mere
cluster of suns and worlds, which to the inhabitants
of the remote regions, that can be reached only
by our telescopes, would seem but a mere luminous
spot, like one which lies near the outermost range
of observation, and appears a fac-simile of our own
—impresses on the mind a feeling of awe, of humi-
lity, and of adoration of that Supreme Being, to
whom worlds, and suns, and systems, are but as the
sand on the sea-shore!

Again, when conducted by our investigations
to the invisible universe around us, the *milky-way*,
and the *fixed stars*, of animal life, which the microscope reveals to us, we are overpowered with the
contemplation of the minutest as well as of the
mightiest of His works! And if, as an eminent
philosopher observes, our planetary system was

* See Mr. Whewell's Bridgewater Essay. The quotation, page
22, should be referred to this admirable work.

gradually evolved from a primeval condition of matter, and contained within itself the elements of each subsequent change, still we must believe, that every physical phenomenon which has taken place, from first to last, has emanated from the will of the Deity.*

With these remarks, I conclude this attempt to combine a condensed view of geological phenomena, with a familiar exposition of the inductions by which the leading principles of the science have been established. And if I have succeeded in explaining, in a satisfactory manner, how by laborious and patient investigation, and the successful application of other branches of natural philosophy, the wonders of geology have been revealed—if I have removed but from one intelligent mind any prejudice against scientific inquiries which may have been excited by those who have neither the relish nor the capacity for philosophical pursuits—if I have been so fortunate as to kindle in the hearts of others that intense and enduring love and admiration for natural knowledge, which I feel in my own,—or have illuminated the mental vision with that intellectual light, which once kindled can never be extinguished, and which reveals to the soul the beauty, and wisdom, and harmony of the works of the Eternal, I shall indeed rejoice, for then my exertions will not have been in vain. And although my humble name may

* Professor Sedgwick.

be soon forgotten, and all record of my labours be
effaced, yet the influence of that knowledge, however feeble it may be, which has emanated from
my researches, will remain for ever ; and, by con-
ducting to new and inexhaustible fields of inquiry,
prove a never-failing source of the most pure and
elevated gratification.   For to one imbued with
a taste for natural science, Nature unfolds her
" hoarded poetry and her hidden spells ; " for him
there is a voice in the winds, and a language
in the waves—and he is

> —— " Even as one,
> Who, by some secret gift of soul or eye,
> In every spot beneath the smiling sun,
> Sees where the *springs of living waters lie !* "
>
> Mrs. Hemans.

Geology, beyond almost every other science, offers fields
of research adapted to all capacities, and to every condition and
circumstance in life in which we may be placed.   For while
some of its phenomena require the highest intellectual powers,
and the greatest acquirements in abstract science, for their successful investigation, many of its problems may be solved by
the most ordinary intellect, and facts replete with the deepest
interest may be gleaned by the most casual observer.

To the medical philosopher Geology presents peculiar attrac-
tions for those hours of leisure and relaxation, which are indispensable to maintain a healthy state of mind ; for it requires
the cultivation and application of chemistry, botany, comparative anatomy, zoology, and physiology—sciences which form
the very foundation of medical knowledge.   It exerts, too,
the most salutary influence, by calling forth the continual
exercise of our intellectual powers ; for the desire to explain
what is obscure in the natural records of the past, induces a

more accurate examination of existing physical phenomena, and of the organization and habits of the living beings within the reach of actual observation. It enforces the necessity of weighing the conflicting evidence of apparently irreconcileable phenomena, of detecting differences or seeking analogies, and of generalizing and combining an immense number of isolated facts. The mind thus acquires the power of acute observation, of patient investigation, and of salutary caution in drawing inferences and arriving at conclusions—habits of the first importance in the discrimination and treatment of diseases. I, therefore, entreat my younger medical brethren, who have favoured me by their attendance on these lectures, not to be deterred from the pursuit of so legitimate a source of the most elevated gratification, by the apprehension lest their professional success should be retarded by a reputed taste for science. Happily the time is now arrived, when the empty boast of possessing *only* professional knowledge, is no longer considered a proof of superior medical skill, but, on the contrary, an unequivocal acknowledgment of limited acquirements, and evidence of contracted and imperfect views of the subjects embraced by medical philosophy. They may rest assured that the practitioner who, by the exercise of his reasoning powers in scientific investigations, is capable of comprehending the laws by which organization and vitality are governed, and is thus enabled, not by mere habit or conjecture, but by cautious induction, to trace the phenomena observable in aberrations of health to those organs on whose functional or structural derangement they may depend, but with which they may appear to hold but obscure or uncertain relations, will ultimately meet a sure reward in the confidence and approval of the unprejudiced and the intelligent. In this, my last public attempt to encourage a taste for scientific pursuits, I may be permitted to allude to my own successful medical career, in proof that the pursuit of science is not incompatible with a deep devotion to professional duties; and I will venture to add, not from vanity or presumption, but from an earnest desire to remove the apprehension which, I know, deters many medical practitioners from pursuits so congenial to their taste and education, that so

far from my known scientific predilections having proved inju-
rious to my professional prospects, they have, on the contrary,
largely contributed towards my success, by affording introduc-
tions which otherwise would not have been within my reach ;
independently of the privilege, which in my estimation is
beyond all price, of communion with the most eminent philosophers of our times.

APPENDIX.

K. Page 431.—On a peculiarity of structure in the
first caudal vertebra of the adult Gavial.—The
discovery of the Swanage crocodile induced me to institute a rigorous examination of the skeletons of the
recent Gavials in the museum of my friend Dr. Grant, of
the London University, with the view of determining the
affinities of the fossil remains. In the course of my investigation I detected a peculiar conformation in the *first
caudal* or *coccygeal* vertebra of the recent gavial, which,
strange to say, appears to have escaped the notice of previous observers. The vertebræ of the existing crocodilian
family are invariably concave in front, and convex behind;
but the *first caudal* (3) in the adult gavial is *doubly convex;*
and the last sacral vertebra (2.) *concave posteriorly*, to
receive the anterior convexity or ball of the caudal. These
peculiarities are shown in the annexed sketch.

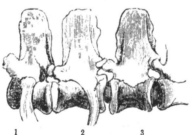

1   2   3

Tab. 77. Sacral and Caudal Vertebræ of the Gavial.

Figs. 1. 2. *The sacral vertebræ.* 3. *The first caudal or coccygeal vertebra, which is doubly convex.*

The last *cervical* vertebra in turtles and tortoises has a similar structure. In a very young gavial in Dr. Grant's collection, the sacro-coccygeal surfaces are as flat as in the vertebræ of mammalia; while in the crocodile and alligator, of the same early period, the coccygeal vertebra is convex in front, as in the adult gavial. This mechanism confers the power of free motion without risk of dislocation or mutilation. The importance of a knowledge of this fact to the palæontologist is too obvious to require remark; the discovery in the Tilgate grit of a caudal vertebra, having both the extremities convex, would, I must confess, have been very perplexing, previously to my examination of the adult gavial.

L. Page 661.—Rev. J. B. Reade, on Fossil Infusoria; in a Letter to Gideon A. Mantell, Esq. LL.D. &c.

My dear Sir,

You are aware that a microscopic examination of recent and fossil plants has not only enabled me to establish some important facts in vegetable physiology, but has also led me to pursue an investigation intimately connected with " the Wonders of Geology." With respect to plants, I have already shown that the solid materials which are contained *in their ashes*, must be ranked among their essential elements; and that while the carbon may be readily dissipated by heat, their solid and earthy ingredients, *whether silica or lime*, so perfectly retain the form and characters of the cells and tubes into which they enter, that the burnt and unburnt specimens have sometimes been mistaken, the one for the other (*see page* 565). I premise this remark, because it enables me to reply to your query, respecting the existence of organic structure in granite, by observing, in the first place, that much of what I have stated with regard to plants, is equally applicable to large portions of the animal kingdom also, and especially to that section of it, viz. the infusoria, which might appear, at first sight, to be wholly removed from such speculations.

My original inquiry having thus conducted me to the con-clusion, *that silicious organization is not destructible by the agency of heat*, I thought it not unreasonable to infer that a careful and more extended microscopic examination into the condition of silica, might lead to the discovery of ele-mentary organic forms, even in the primitive strata them-selves. It was obviously not necessary to exclude granite from this examination, under the common and apparently natural impression, that the igneous fusion which preceded the present arrangement of its particles, would destroy every trace of organization; for I had before me too many manifest proofs, that an intense white heat, though capable of fusing glass, was incapable of effecting any change in the minute silicious organization both of plants and animals. Moreover, there appeared to be a strong suspi-cion in some minds, that every successive surface of our globe had been characterized by its own minute living forms; and you, yourself, had more than once contended for the existence of life during the granitic period. To give a reality, however, to a *first condition*, thus pronounced to be *probable*, we must discover the skeletons of animalcules even in granite itself. But here arises a difficulty which it will baffle our utmost ingenuity to remove; for, though, on the one hand, I meet with silicious corpuscles in the primitive rocks, and find, on the other hand, that the in-destructible organic skeletons of recent infusoria exhibit, even under a power of 900 linear, a striking similarity of form, yet the entire absence of external structure precludes me from assigning a common animal origin to the ancient and recent organisms. Still, the inquiry, even in its pre-sent state, is far from being fruitless; for it cannot but be a matter of surprise, that immense mountain masses should have been found to consist of an aggregation of symmetrical bodies, between $\frac{1}{5,000}$ and $\frac{1}{10,000}$ th of an inch in diameter, articulated together in the form of rings, as in chalk, or of slender threads, as in limestone, and the quartz of granite, and that an exact counterpart of this curious structure in the mineral kingdom should be exhibited in the vegetable, by the mouldiness of paste, and in the animal by the *Gaillo-nella ferruginea.*

The " Philosophical Magazine" for April, 1837, to which

686   APPENDIX.

my attention has been recently directed, contains, under the article "Palæontology," a short note by Professor Ehrenberg, on the organic forms which he had observed during an exact microscopic analysis, several times repeated, of upwards of a hundred minerals of different groups; and in the valuable papers of the same author, published originally in "Poggendorf's Annalen," and lately given to the English reader in "Scientific Memoirs," vol. I. part iii. it is observed, with respect to *Gallionellæ*, and other species of animalcules, that their proportion merits a passing attention. "The millions of the tribes of infusoria have often been mentioned and spoken of, almost without consideration of their number, perhaps because little belief is entertained of their corporeality. But since the Poleirschiefer of Bilin must be acknowledged to consist almost entirely of an aggregation of infusoria, in widely extended layers, without any connecting medium, these animals begin to acquire a greater importance, not only for science, but for mankind at large. A cubic inch of the Poleirschiefer would contain, on an average, about 41,000 millions of the *Gaillonellæ*; and the silicious shield of each animalcule weighs about the 1-187 millionth part of a grain."

Explanation of the Figures, (Tab. 78, 79, 80,) drawn under a magnifying power of about 500 linear.

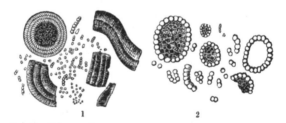

1   2

TAB. 78.—BODIES IN PORCELAIN EARTH AND CHALK *(highly magnified.)*

Fig. 1. Form of porcelain earth, exhibiting concentric articulated rings; entire and in fragments.

Fig. 2. Chalk: the elementary molecules articulated in the form of rings; entire and in fragments.

TAB. 79.—CIRCULAR BODIES IN MICA, *(highly magnified.)*

Flat circular bodies imbedded in Mica, corresponding in size and appearance with the rings of *Gaillonella distans*, and owing to their peculiar action upon light, presumed to be silicious. Professor Ehrenberg observes, that "Mica and Quartz present *a granulated appearance of great regularity*, either without their outer surface of fracture undergoing any previous preparation, or after having been warmed or heated to redness."

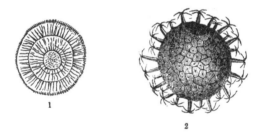

1   2

TAB. 80.—INFUSORIA IN FLINT, *(highly magnified.)*

Fig 1. *Body unknown—formed of three distinct circles: the intervening spaces are filled with numerous delicate rays, and the exterior circle is sinuous and fringed. From the flint of Sydenham in Kent.*

2. *Supposed by M. Turpin to be the egg of a Polype* (Christatelle vagabonde), *but is the* Xanthidium furcatum *of Ehrenberg. This fossil animalcule occurs with the former in the flint of Sydenham.*

An accurate microscopic examination of flint nodules would lead to the discovery of a considerable number of hitherto unknown organic forms.

We may regard as hitherto ascertained facts, concludes Professor Ehrenberg, that

1. Bergmehl......................................⎫
2. Kieselguhr ....⎬ Newest formation,
3. Poliershiefer ....⎫
4. Saugschiefer ....⎬ Tertiary formations,
5. The semi-opal of the Polierschiefer ...⎭

consist entirely, or partly, of the shells of shield-infu-
soria.

The following species of stone are *very probably* of the
same nature :

6. The semi-opal of the Dolerite............⎫
7. The (precious) opal of the Porphyry...⎬ Secondary and primary
8. The flint of the Chalk ....⎭   formations.

It is gratifying to know that Professor Ehrenberg is still
engaged in a close examination of the remarkable characters of the primary formations, and that he has announced
his intention to publish the results, when sufficiently ma-
tured.

Believe me to be, my dear Sir,

Very faithfully yours,

J. B. READE.

PECKHAM,
      Dec. 1837.

EHRENBERG ON INFUSORIA.—To the above interesting
communication of Mr. Reade, I will add the following
abstract of Ehrenberg's observations on the same subject.
     This eminent observer has determined twenty-eight fossil
species of infusoria, all belonging to the family of the
Bacillariæ. Of these, fourteen species are undistinguish-
able from existing fresh-water, and five species from marine infusoria; the others belong to extinct or unknown
forms. The great sharpness of the outlines of all these
silicious shields appears to have been produced by intense
heat, by which all organic (particularly vegetable) carbon has
been dissipated; for some of these animals, like existing
species, must have lived on plants. Different kinds of the
minerals containing the fossil infusoria have a preponderance
of different species. The polishing slate or tripoli of Bilin,
consists almost entirely of an aggregation of infusoria in

layers, without any connecting medium; and of this stone about 50 cwt. are consumed annually at Berlin. The size of a single specimen of these infusoria is equal to 1-6th of the thickness of a human hair: as the stone is slaty, but without cavities, the animalcules lie closely compressed. About 23 millions of these creatures would make up a cubic line; and in a cubic inch there would be 41,000 millions, weighing 220 grains; the silicious shield of each animalcule weighs about 1-187 millionth part of a grain. The fossil animalcule of the iron ochre is only the 1-21st part of the thickness of a human hair; and one cubic inch of this ochre must contain *one billion* of the skeletons of living beings!

The *infusoria* rock of Bilin forms a bed fourteen feet thick, above a layer of clay which rests on chalk marl, beneath which are primary rocks. The upper beds of stone rest on a projected mass of basalt, which forms the Spitalberg; on the opposite side of which coarse limestone, with many crinoidea and other chalk fossils, lies on the gneiss. The harder masses, containing semi-opal, are situated in the upper part of the tripoli. A close microscopical analysis of the semi-opals from Bilin, which equal flint in hardness, shows that it consists partly of infusorial forms, held together by a small quantity of transparent silicious cement, and partly of single infusoria, but of a larger size, like insects in amber. From the power possessed by these animalcules, of secreting skeletons of iron, flint, and lime, the proverb, *Omnis calx e vermibus, omnis silex e vermibus, omne ferrum e vermibus,* seems likely to be verified in a very striking manner.—*Taylor's Scientific Memoirs, Vol. I. Part 3.*

Recent Zoophytes.

Fig.

1. Corallium rubrum, from the Red Sea; p. 477.
   a a. The animal tissue with the polypi expanded.
   b. The red calcareous axis or skeleton, to which the term *coral* is popularly applied.
2. The polype of a species of Flustra, highly magnified.
   a. The arms; p. 475.
3. *Fungia rubra*, as seen alive in the sea; from the Ladrone islands; 1-12th the natural size; p. 482.
4. A branch of Campanularia, highly magnified. One of the polypi is seen in its cell; in the other the tentacula are protruded; p. 475.
5. *Madrepora Corimbosa*, from the Indian Ocean; as seen alive in the water; reduced to 1-4th the size of the original; p. 479.
6. A branch of *Corallium rubrum*, with its fleshy investment, and the polypi expanded. Drawn from a living specimen; p. 477.
7. *Flustra bombycina*, a portion highly magnified to show the cells; p. 473.
8. *Gorgonia patula;* magnified view of a branch, with two polypi; one in an expanded, and the other in a contracted state; p. 475.
   a. The dark-coloured central axis; b. a polype expanded.
9. *Pocillopora cerulea*, as seen alive; from the Indian Seas; p. 482.
10. *Halina papillaris*, a spreading sessile species of sponge, or porifera, from the shore at Brighton; p. 458.
10a 10b 10c. Spines of porifera; p. 495.
11. *Meandrina cerebriformis*, or brain coral, as seen alive in the water from the American Sea. Reduced 1-30th; p. 484.
12. *Flustra bombycina* attached to a fucus; natural size; p. 473.
13. Caryophillia angulosa, from the American Seas; reduced one-half; p. 482.

DESCRIPTION OF PLATE II.

Recent Zoophytes.

Fig.

1. *Pavonia lactuca;* a single cell, inclosing a polype; from the shores of the South Sea Islands; p. 483.
   a. The delicate calcareous foliated expansions of the coral.
   b. The oral disc, with its marginal tentacula.

2. Branch of a Gorgonia from the West Indies, presented by Miss Emily Lindo; p. 475.

3. *Fungia patelliformis*, from the Mediterranean, presented by Miss Crofts; the calcareous skeleton of a coral; p. 482.

4. Gorgonia, from the Mediterranean: at the base the dark central axis is seen; p. 476.

5. *Fungia actiniformis*, from the South Pacific Ocean; as seen alive, and the polypi extended; p. 482. From Dr. Grant's Comparative Anatomy.

6. *Sertularia setacea*, from the sea off Brighton. From Mr. Lister's Memoir; p. 474.

7. *Astrea viridis*, as seen alive, with some of the polypi expanded; p. 483.
   a. The mouth. b. The stem, or body of the polype. c. The tentacula. f. The polypi contracted within the cells. g h. The polygonal cells and calcareous base of the lithophyte, the fleshy covering being removed. From Dr. Grant's Comparative Anatomy.

7a. A polype of *Astrea viridis*, highly magnified. a. The mouth. b. Annular striæ. c. The tentacula. d. The striated body or stem of the polype.

8. *Sertularia pinaster*, alive; p. 474.

8a. Branch of *Sertularia pinaster*, highly magnified.

9. *Flustra pilosa;* a portion very highly magnified; the polypi are seen partly protruded from their cells; p. 461.

10. *Sarcinula musicalis*, or organ-pipe coral, from the coast of New South Wales, as it appears when alive in the water, and the beautiful green polypi expanded; p. 478.

10a. Three connected tubes of *Sarcinula musicalis*, magnified, showing the interior of the tubes; p. 478.

11. *Turbinolia cyathus*, from the Mediterranean; the calcareous axis or skeleton; p. 481.

12. A group of Actiniæ, as seen alive in the water; p. 480.

12a. Magnified view of one of the tentacula.

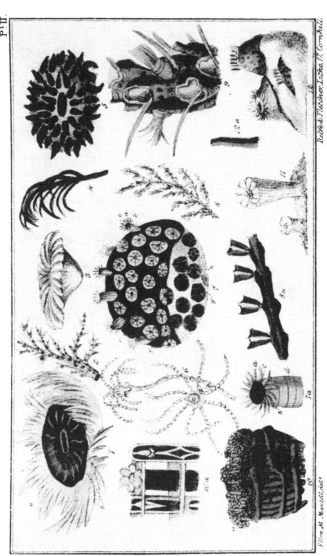

ZOOPHYTES.

F:In M. Mantell, del.°   Pub'd by Fletcher. Litho.º. H. Corbould.

Pl.3.

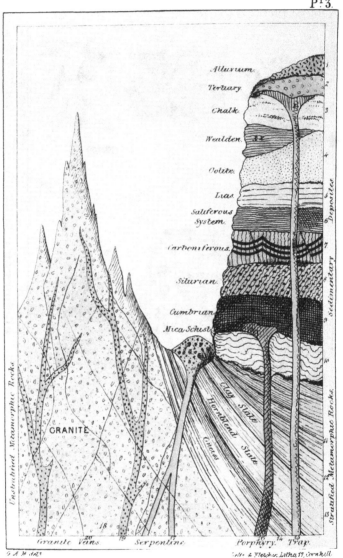

Alluvium.

Tertiary.

Chalk.

Wealden.

Oolite.

Lias.

Saliferous System.

Carboniferous.

Silurian.

Cumbrian.

Mica Schist.

Sedimentary Deposites.

Unstratified Metamorphic Rocks.

GRANITE

Clay Slate

Hornblend Slate

Gneiss

Stratified Metamorphic Rocks.

Granite Veins. Serpentine. Porphyry. Trap.

G. A. M. del.?

Colls & Fletcher, Litho, 17, Cornhill.

DESCRIPTION OF PLATE III.

Theoretical Arrangement of the Rocks which compose the Crust of the Earth; p. 177.

1. Alluvial deposites.
2. Tertiary formation.
   Cretaceous system of strata, comprising the
     Chalk, with and without flints,
     Chalk marl,
     Galt,
     Green Sand.
3*.The Wealden.
4. The Oolite.
5. Lias.
6. Saliferous system, comprising,
     1. The New red sandstone,
     2. Magnesian limestone.
7. The Carboniferous system, namely,
     The Coal-measures,
     The Mountain, or Carboniferous limestone,
     The Old red sandstone,
8. The Silurian system of Mr. Murchison ; consisting of the Old red sandstone, Ludlow rocks, Dudley limestone, Caradoc sandstone, Landeilo flags, Trilobite slate.
9. The Cumbrian, or Cambrian system of Professor Sedgwick ; consisting of the series of greywacke and slate rocks.
10. Mica slate series.
11. Conglomerates.
12. Clay slate.
13. Hornblende slate.
14. Gneiss.
15. Dyke of Trap.
16. Dyke of Porphyry.
17. Dyke of Serpentine, &c.
18. Granite.
19. Granite veins.
20. Metalliferous veins.

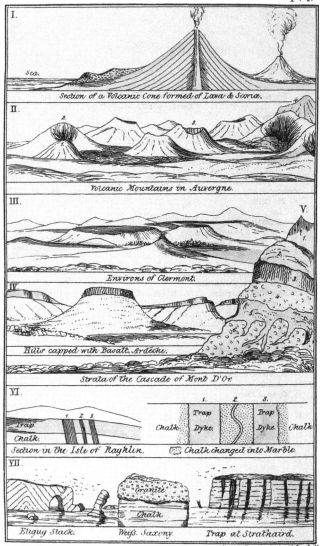

I.

Sea.

Section of a Volcanic Cone formed of Lava & Scoriæ.

II.

Volcanic Mountains in Auvergne.

III.

V.

Environs of Clermont.

IV.

Hills capped with Basalt, Ardèche.

Strata of the Cascade of Mont D'Or.

VI.

Trap.

Chalk.

Section in the Isle of Raghlin.

Chalk	Trap Dyke.		Trap Dyke.	Chalk
1.		2.	3.	

Chalk changed into Marble.

VII.

Granite

Chalk

Eligug Stack.   Weyß. Saxony.   Trap at Strathaird.

G. A. M. del⁰.   Reife & Fletcher; Litho, 11, Cornhill.

DESCRIPTION OF PLATE IV.

I. Section of a volcanic cone, formed of scoriæ and lava. A bed of alluvial detritus (coloured *sienna*) covers the flanks of the principal cone, as in the Isle of Ascension, Iceland, &c.

II. Volcanic mountains in Auvergne, from Mr. Scrope. Part of the southern chain of Puys, exhibiting the broken craters of Chaumont, each with a lava current issuing from the base; see p. 240.
1. Montchal.
2. Puys de Montgy.
3. Montjughat.
4. Mont Dome, in the distance.

III. Environs of Clermont; from Mr. Scrope; *vide* p. 240. The town of Clermont is seen in the valley; in front is a basaltic peak, on which is built the castle of Montrognon. The green on the hills beyond denotes the basaltic platform, which caps hills of freshwater limestone. The distant outline is the granitic escarpment forming part of the boundary of the plain of Auvergne.

IV. Hills capped with basalt; Ardêche. View of the lateral embranchments of the basaltic platform of the Coiron, in Ardêche; see p. 241.
The beds of basalt, between three and four hundred feet in thickness, are spread over limestone strata, which, together with the basalt, were once continuous, but have been eroded and carried away by alluvial action.

V. Section of the cascade of Mont Dor; see p. 243. 1. Porphyritic trachyte; *volcanic*. 2. Tufa; a deposit from fresh-water. 3. Basaltic phonolite 4. Breccia, composed of volcanic fragments. 5. Basalt. 6. Tufa, with veins of basalt.

VI. Section in the Isle of Rathlin. Eruption of trap through chalk; p. 646.
Figs. 1, 3. Trap dykes: the chalk between the dykes, and on each side of the walls, to an extent of several feet, is changed into granular marble.
Fig. 2. A vein of trap traversing altered chalk.

VII. Eligug* Stack: a range of cliffs, composed of carboniferous limestone, in which the strata have been contorted by elevatory movements, and the upper part removed by denudation; from Mr. De la Beche.
Granite on Chalk, near Weiss, in Saxony. This section shows that the metamorphic rock has been erupted since the deposition of the chalk, and has flowed over the cretaceous strata on which it reposes.
Trap with sandstone, at Strathaird. Vertical dykes of trap intersecting horizontal strata of sandstone; p. 647.

---

* Eligug—so called from the number of sea fowls, principally the Eligug (*arca torda*), which frequent it.

Pl. 5.

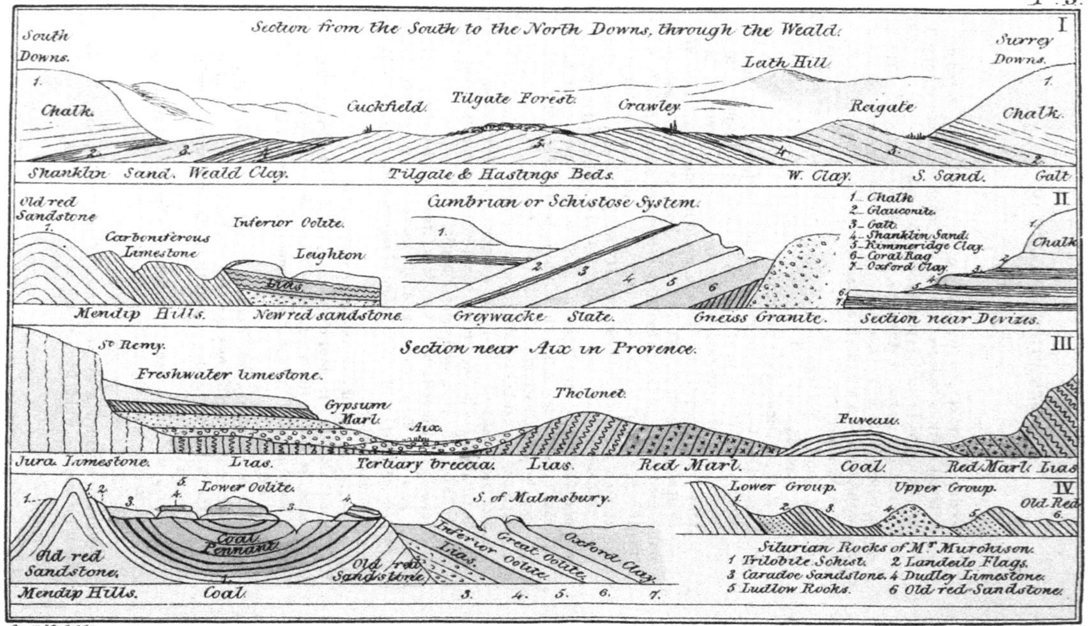

I

Section from the South to the North Downs, through the Weald.

South
Downs.

Chalk.

Surrey
Downs.

Chalk.

Cuckfield. Tilgate Forest. Crawley. Reigate. Leith Hill.

Shanklin Sand. Weald Clay. Tilgate & Hastings Beds. W. Clay. S. Sand. Galt.

Old red
Sandstone.

Carboniferous
Limestone.

Inferior Oolite.

Leighton.

Bala.

New red sandstone.

Greywacke.

Slate.

Gneiss Granite.

Cambrian or Schistose System.

Mendip Hills.

II

1. Chalk.
2. Glauconite.
3. Galt.
4. Shanklin Sand.
5. Kemmeridge Clay.
6. Coralline
7. Oxford Clay.

Chalk.

Section near Devizes.

III

Section near Aix in Provence.

St Remy. Freshwater limestone. Gypsum Marls. Aix. Tholonet. Fuveau.

Lias. Tertiary breccia. Red Marl. Coal.

Jura limestone.

IV

Lias. Red Marl. Lias. Lower Group. Upper Group. Old. Red.

Mendip Hills. Coal. S. of Malmsbury. Old red
(Sandstone) Inferior Oolite. Great Oolite. Oxford Clay.

Lower Oolite.

1. 2. 3. 4. 5. 6. 7.

Silurian Rocks of Mr Murchison.
1 Trilobite Schist. 2 Llandeilo Flags.
3 Caradoc Sandstone. 4 Dudley Limestone.
5 Ludlow Rocks. 6 Old red Sandstone.

Reeve & Fletcher lithos 17, Lombard

V. A. Malby & Sons

DESCRIPTION OF PLATE V.

I. Section from the South to the North Downs, through the Weald of Sussex; see page 317. 1. Upper and lower white chalk, and chalk marl. 2. Galt. 3. Shanklin sand. 4. Weald clay. 5. Tilgate and Hastings strata.

II. Section of the Mendip Hills; see page 524. 1. Old red sandstone. 2. Carboniferous limestone. 3. New red sandstone. 4. Lias. 5. Inferior Oolite.

Cumbrian or Schistose System of Professor Sedgwick, page 608; from Pl. 1, Encyclop. Metrop. 1. Carboniferous System. 2. Greywacke slate, with shells. 2*. Limestone with corals and shells. 3. Green slate. 4. Red argillaceous rock and dark clay slate. 5. Chiastolite slate. 6. Hornblende slate. 7. Gneiss. 8. Granite.

Section near Devizes; see page 391. 1. Chalk. 2. Glauconite, or Firestone. 3. Galt. 4. Shanklin sand. 5. Kimmeridge clay. 6. Coral rag. 7. Oxford clay.

III. Section near Aix in Provence, by Mr. Lyell and Mr. Murchison; see page 228. 1. Jura limestone—equivalent to the Oolite. Tertiary blue limestones and marls: *erroneously* marked *Lias* in the plate. 2. Tertiary red marls. 3. Tertiary coal. 4. The tertiary beds repose unconformably on the Jura limestone, and consist, on the north of the Valley of Aix (the left side of the section of):—1. Breccia or conglomerate. 2. Foliated marls. 3. Gypsum. 4. Freshwater limestone. On the south, towards Fuveau, red marls, limestones, and shale, with coal fit for fuel, are disposed as in the section.

IV. Coal basin of Somersetshire; see page 524; from Mr. Conybeare. Old red sandstone of the Mendip Hills. 1. Carboniferous limestone. 2. Millstone grit. Coal—and Pennant grit. 3. New red sandstone. 4. Lias. 5. Inferior Oolite. 6. Great Oolite. 7. Oxford clay—south of Malmesbury; see page 391.

Silurian System of Mr. Murchison; page 605. 1. Trilobite slate. 2. Landeilo flags. 3. Caradoc sandstone. 4. Wenlock (Dudley) limestone. 5. Ludlow rocks. 6. Old red sandstone.

Pl.6.

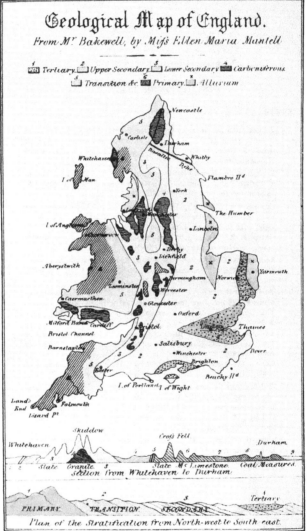

Geological Map of England.

From Mr Bakewell, by Miss Ellen Maria Mantell

1 Tertiary. 2 Upper Secondary. 3 Lower Secondary. 4 Carboniferous.
5 Transition &c. 6 Primary. Alluvium.

Section from Whitehaven to Durham.

Plan of the Stratification from North-west to South-east.

DESCRIPTION OF PLATE VI.

———

A Geological Map of England, reduced from Mr. Bakewell's, and slightly
modified. It serves to illustrate the general distributions of the various
groups of strata or formations over England : namely—

    1. Tertiary.
    2. Upper Secondary.
    3. Lower Secondary.
    4. Carboniferous beds.
    5. The Transition, or Silurian System.
    6. The Cumbrian and metamorphic, or primary rocks.
    7. Alluvial deposites.

Section from Whitehaven to Durham, or from the Irish Sea, through
Cumberland, to the North Sea; from Mr. Conybeare. It exhibits the
succession of the formations from the Magnesian limestone to the Slate
rocks; and the disruptions which have taken place in the strata of
that part of England. 1. Magnesian limestone. 2. Coal and mountain
limestone. 3. Slate. 4. Granite. 5. Broken coal and greenstone.
6. Mountain limestone. 7. Millstone grit. 8. Coal measures. 9.
Magnesian limestone.

Plan of the Stratification of England from North-west to North-east.

    1. Primary, or unstratified metamorphic rocks.
    2. Stratified metamorphic rocks, or Transition series.
    3. Secondary formations.
    4. The Tertiary.

DESCRIPTION OF THE ILLUSTRATIVE WOOD ENGRAVINGS.*

VOLUME I.

* The Wood Engravings are by Mr. Wheeler, 14, Calthorpe-street, Gray's Inn-lane.

Y Y 4

Description of Wood Engravings

INDEX.

A.

Y Y 5

INDEX.

INDEX.

THE END.

R. CLAY, PRINTER, BREAD-STREET-HILL.

CORRIGENDA.

The pressure of professional engagements having prevented a careful revision of the work by the author, some errors have escaped notice, and a few references to other writers have been omitted; some of the Errata are noticed in the following list.

Page 5, line 1. for *called upon* read *required to.*
35,   24, for *rivers* read *river.*
49,   2, for *fusible* read *soluble.*
75,   21, for *marbles* read *masses.*
78.   8, for *source* read *foci.*
99,   6, for *in* read *beneath.*
223.   11, for *enumerated* read *enunciated.*
248,   22, for *water* read *crater.*
313,   2, for *genus* read *genus of reptiles.*
317,   3, for *Reigate-hill* read *Cockshut-hill.*
346,   24, for *of reptiles, bones, &c.* read *of bones of reptiles.*
406,   29, for *cube-spar* read *calc-spar.*
498,   10, for *circinnatis* read *circinnalis.*
359, (the bottom.) *Appendix.* Add :—" This specimen was purchased of Mr. Bensted in its broken state, and presented to Dr. Mantell by the following individuals, the proposition originating with the two gentlemen whose names stand first on the list, viz.—Horace Smith, Moses Ricardo, Thomas Attree, George Basevi, Thomas Bodley, R. Heaviside, E. Lindo, J. J. Masquirier, W. Tenant, and T. Saul, Esqrs. ; Drs. Hall and Price ; Revds. J. S. M. Anderson, Thomas Rooper, and H. M. Wagner ; and Sir Richard Hunter."

Printed in the United States
By Bookmasters